Proceedings of the Symposium of Aeronautical and Aerospace Processes, Materials and Industrial Applications

P. Zambrano-Robledo
A. Salinas-Rodriguez
F. Almeraya Calderon
Editors

Proceedings of the Symposium of Aeronautical and Aerospace Processes, Materials and Industrial Applications

Presented at the XXV International Materials Research Congress – Cancun 2016

Springer

Editors
P. Zambrano-Robledo
Universidad Autónoma de Nuevo León
Facultad de Ingeniería Mecánica y Eléctrica
San Nicolás de los Garza, NL, Mexico

A. Salinas-Rodriguez
CINVESTAV Unidad Saltillo
Ramos Arizpe, Coah, Mexico

F. Almeraya Calderon
Universidad Autónoma de Nuevo León
Facultad de Ingeniería Mecánica y Eléctrica
San Nicolás de los Garza, NL, Mexico

ISBN 978-3-319-65610-6 ISBN 978-3-319-65611-3 (eBook)
DOI 10.1007/978-3-319-65611-3

Library of Congress Control Number: 2017954935

Printed on acid-free paper

This Springer imprint is published by Springer Nature
The registered company is Springer International Publishing AG
The registered company address is: Gewerbestrasse 11, 6330 Cham, Switzerland

Contents

Part I
Microstructure and Characterization

Microstructural Characterization of Inconel 718 for Future Ultrasonic Fatigue Testing

Manuel Guzmán-Tapia, Gonzalo Mariano Domínguez Almaraz, Ishvari Fernanda Zúñiga Tello, Alberto Gutiérrez Martínez, and Julio Cesar Verduzco Juárez

Abstract This work presents the effects of the solid solution and double aging heat treatments of the superalloy Inconel 718 (I-718) and the characterization of their microstructure properties by optical microscope (OM), scanning electronic microscope (SEM), hardness Vickers (HV), and hardness (RA, RB, and RC). The relationship between microstructure and mechanical properties, according to the norm AMS5663 (Aerospace Material Standards, USA), was studied too. Solid solution heat treatments were performed at 1078 °C/1 h and water quenching. Double aging heat treatment was carried out at 720 °C/8 h and then cooling furnace 1 h to 620 °C/8 h and then cooling by air. Solid solution heat treatment dissolved precipitates, and boundary grains and δ phase (Ni_3/Nb) also reduced hardness which facilitates specimens machining with dimensions and morphology; these dimensions will be used for future ultrasonic fatigue testing and have been obtained according to a modal analysis by finite element method. Once specimen machining was completed, double aging heat treatment (precipitation hardening) was performed; this heat treatment leads to increased hardness according to the norm AMS5663 for subsequent tests in ultrasonic fatigue.

Keywords Inconel 718 • Microstructural characterization • Heating treatments

1 Introduction

Inconel 718 is Ni-based superalloy, commonly known as Ni-Fe based, has been developed as a structural material in hot sections of jet engines such as turbine disks, blades, etc., because of its excellent mechanical properties at high temperatures and

M. Guzmán-Tapia (✉) • G.M. Domínguez Almaraz • I.F. Zúñiga Tello
A. Gutiérrez Martínez • J.C. Verduzco Juárez
Posgrado en Ciencias en Ingeniería Mecánica U. M. S. N. H,
Av. Francisco J. Múgica S/N C. P, 58140 Morelia, Michoacán, Mexico
e-mail: manolo_verrckt@yahoo.com.mx

P. Zambrano-Robledo et al. (eds.), *Proceedings of the Symposium of Aeronautical and Aerospace Processes, Materials and Industrial Applications*,
DOI 10.1007/978-3-319-65611-3_1

good corrosion resistance and it can be heat treated to different strength levels. Turbine blades suffer several operation conditions during a flight, such as (1) operation environment (high temperature, fuel and air contamination, solid particles), (2) high mechanical stresses (vibration, centrifugal and aerodynamic forces), and (3) high thermal stresses due to thermal gradients [1, 2]. The major failure modes reported are mechanical degradation, including thermal fatigue, thermal mechanical fatigue, low cycle fatigue, and creep, and environmental damage such as high temperature, oxidation, hot corrosion, and erosion. The combination of these two modalities of degradation is also considered [3–6].

It is known that hot corrosion is a degradation process of material components, due to the chemical or electrochemical effect of a certain contaminants, such as NaCl, Na_2SO_4, and V_2O_5. In order to identify the kind of corrosion, the modalities are classified depending on the temperature: (1) high-temperature hot corrosion and (2) low-temperature hot corrosion [7, 8].

Inconel 718 is an age hardening alloy, in which the main phases for solid solution and aging heat treatment are γ (Ni or Cr with FCC structure) for the matrix, γ' (Ni_3 (Al/Ti) with FCC structure), and γ'' (Ni_3Nb with BCT structure). Both $\gamma'$and γ'' are precipitate phases for strengthening, whereas δ phase (Ni_3Nb with orthorhombic structure) does not contribute to the hardening, but its presence implies a loss of hardenability of the alloy and metal carbides (MC), the last ones contributing effectively to increase the hardness of the matrix γ'. In the MC, M could be Cr, Ti, Mo, W, Nb, Hf, and Ta [9].

There are two current factors to improve a superalloy performance: (1) changing elements of the composition, for example, addition of Al, Ti, Nb, and Ta, which are used to produce and optimize volume fraction of the γ' or strengthening phase, whereas carbon is intentionally added to generate carbides in order to reduce creep by inhibiting grain boundary sliding, and (2) using appropriate heat treatments (solid solution and aging heat treatments). Solid solution treatment has been used to dissolve or promote phases, to obtain a desire grain size and to eliminate residual stresses; on the other hand, melting temperature point of γ' is 1050 °C, and in order to obtain a desired grain size and strengthening phase, the solid solution heat treatment needs to heat above 1050 °C. Aging treatment was used to precipitate γ' phase [10].

Solid solution and aging heat treatments are the best options for the strengthening mechanism for Inconel 718 by heat treatments [11]: (1) solid solution at 980 °C by 1 h and water quenching and (2) double aging at 720 °C/8 h, furnace cooling (FC) at 620 °C by 1 h and keeping at 620 °C by 8 h, and, finally, air cooling. The aging treatment has the purpose to limit hardness below 40 Rockwell C [12] or 430 HV [13], to prevent stress corrosion cracking as well as sulfide corrosion cracking and/or hydrogen embrittlement. Nevertheless, solid solution couldn't solve carbide and lave phases under 1032 °C, for this nickel alloy. The correct heat treatment is required to ensure proper rate between γ'' and δ phases; the effect of time and temperature on microstructure and mechanical properties has been determined by microstructure changes and measurement of hardness. Temperature of solid solution must be high enough to dissolve alloying elements in metal matrix during aging which generate precipitates rather than harden the nickel matrix at 960 °C, but some amount of δ phase may still be present [14–16].

Yeh et al. [17] have studied the effect on the microstructure and morphology of grain boundaries, and they have found that it could be a function of heat treatment conditions and improvement in creep rupture life: the heat treatment used in this study comprised a solid solution treatment at 1095 °C/1 h, FC at 850 °C/4 h, and 1095 °C/1 h, 850 °C/4 h. On the other hand, Kuo et al. [18] studied the aging effects of the δ phase using three different heat treatments:

1. Solid solution at 1095 °C, 1 h/air cooling (AC), then aging at 955 °C, 1 h/AC at 720 °C, 8 h/FC 57 K/1 h to 620 °C, 8 h/AC.
2. Solid solution at 1095 °C, 1 h/AC, then aging at 720 °C, 8 h/FC 57 K/h to 620 °C; in this heat treatment, the stress rupture life and creep are the largest in comparison with the other heat treatments.
3. Solid solution at 1095 °C, 1 h/AC, then aging at 955 °C, 3.5 h/AC + 720 °C, 8 h/FC 57 K/h to 620 °C, 8 h/AC.

The purpose of the present study is to investigate microstructural changes and the effect on mechanical properties by heat treatments such as (1) solid solution and (2) double aging heat associated with the mechanical property of hardness (Rockwell A, B, and C and microhardness Vickers), in order to propose an alternative sequence of solid solution heat treatment for the nickel alloys I-718.

2 Experimental Procedure

A round bar of Inconel 718 with the dimensions 1/2 inches of diameter and 2.5 inches of length was bought, in which the chemical composition in weight is shown in Table 1.

Solid solution and aging treatments are used to get the benefits of the heat treatments effects. The solid solution usually is the first step because it recrystallizes, homogenizes, and dissolves phases in the matrix structure (FCC), dissolving carbides in grain boundaries [19]. The solid solution treatment was performed at 1078 °C for 1 h; afterward, water quenching is applied as shown in Fig. 1 in order to propose an alternative sequence of solid solution heat treatment for the nickel alloys I-718 for ultrasonic fatigue testing. Working temperature for solid solution has been reported above 1032 °C; this temperature is declared to dissolve carbide and lave phases in the matrix [9, 13, 14].

Double aging treatment has the purpose to increase the strength through precipitation of additional quantities of second phases (γ' and γ'') from the matrix, developed from the solid solution treatment. Double aging treatment was carried out at 720 °C for 8 h, followed by furnace cooling to 620 °C turning down to 10 °C every 6 min. Second aging was performed at 620 °C for 8 h and, finally, air cooling at room temperature (Fig. 2) [18, 19]. The solid solution and double aging were conducted using a furnace Thermolyne model F48020–80.

Table 1 Chemical composition of Inconel 718 (wt %)

Element	Ni	Fe	Cr	Nb	Mo	Ti	Al	Co	Mn	Si	Cu
WT %	52.59	18.7	18.46	5.13	2.89	0.94	0.62	0.22	0.07	0.07	0.05
Element	C	Mg	Ta	P	B	Ca	S	Se	Pb	Bi	–
WT %	0.04	0.01	0.01	0.008	0.004	0.001	0.0005	0.0003	0.0003	0.00003	–

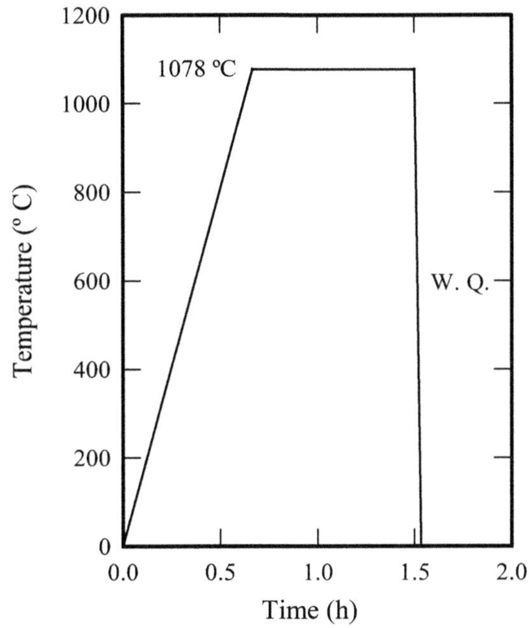

Fig. 1 Diagram of the solid solution treatment with water quenched (WQ)

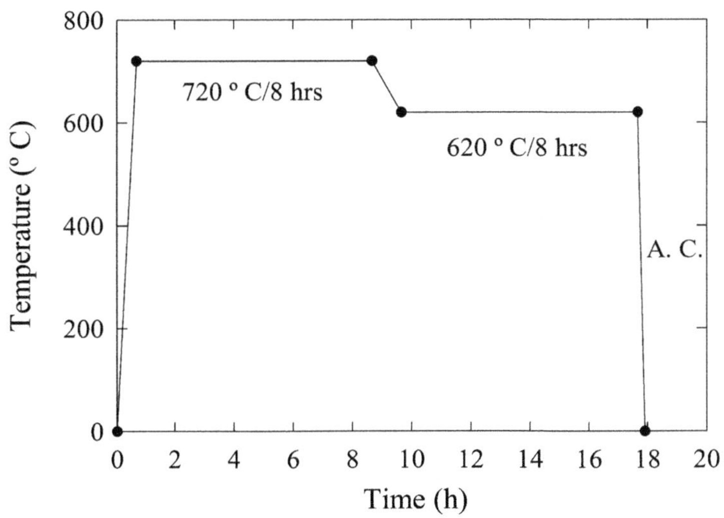

Fig. 2 Diagram of the double aging treatment

Specimen's surfaces were prepared using standard mechanical polishing procedures: first, a piece of the specimen was cut using a Metkon machine model Servocut 301-MA, and then cross sections were mounted in Bakelite in a Metkon machine model Ecopress 100 at 180 °C and 210 Pa for 4 min and then cooled for 3 min, and for grinding it used a SiC grit papers (120, 240, 360, 400, 600, 1000, and

2000). The samples were polished in the same machine on a cloth using 0.1 μm diamond suspension and 0.1 μm alumina suspension as a polishing solution to produce a mirror finish in a Metkon machine Forcipol model 2 V. The etching was carried out with a 50% hydrogen peroxide and 50% hydrochloric acid.

3 Results and Discussion

The microstructure of the superalloy I-718 was analyzed by optical microscope (OM) Jeol model 5910LV. Figure 3 shows images of the matrix metal of as-received I-718 with an annealed heat treatment: the grain size average of γ matrix is about 50 μm according to ASTM E 112 norme, corresponding to a grain size number 6 in this norme; the size of carbides is close to 10 μm as shown in the same figure by the black spots. The grain size in these superalloys should be medium in size (40–50 μm); if the grains are too small, they tend to decrease fracture strength, whereas if they are excessively large, they can damage tensile properties and creep resistance [16].

The microstructure of the superalloy I-718 was analyzed by scanning electronic microscope (SEM) Jeol model JSM-5910LV. On SEM images, Fig. 4a, b, it is appreciated that the most abundant delta phase (δ) is equally distributed into the grain and in the grain boundaries. Some of carbides are also observed such as NbC type detected by EDS analysis in different shapes and sizes (Fig. 5); furthermore, in a EDS analysis, it was found that the presence of prominent Nb peak indicates predominance of niobium carbides over the titanium or boron carbides. Carbon has been detected by EDS analysis in the form of smaller precipitates in the intergranular regions. In literature, it is well known that NbC carbides and δ (Ni_3Nb) have needlelike, globular, amorphous, and blocky morphologies [20].

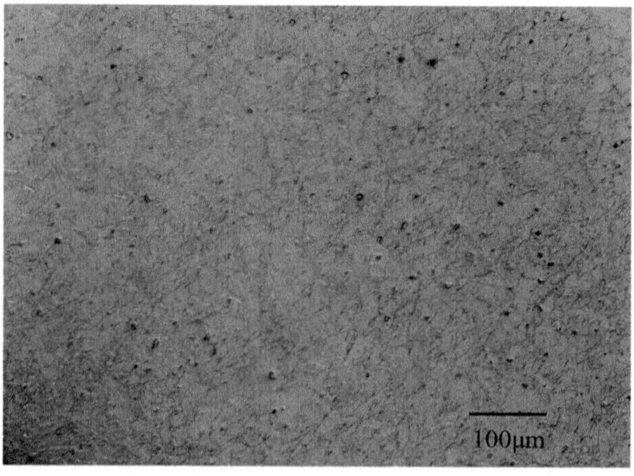

Fig. 3 Optical microscope image of Inconel 718 as-received

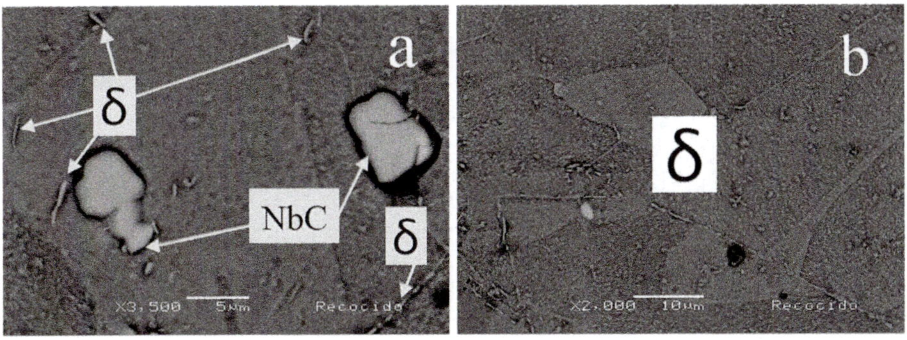

Fig. 4 SEM images of Inconel 718 as-received: (**a**) enhancing NbC carbides and delta phase, (**b**) δ phase inside and in the boundary of grains

Project 1 10/03/2016 13:12:55

Spectrum processing :
No peaks omitted

Processing option : All elements analyzed (Normalised)
Number of iterations = 4

Standard :
B B 20-Jul-1999 12:00 AM
C CaCO3 1-Jun-1999 12:00 AM
Ti Ti 1-Jun-1999 12:00 AM
Cr Cr 1-Jun-1999 12:00 AM
Ni Ni 1-Jun-1999 12:00 AM
Nb Nb 1-Jun-1999 12:00 AM

Element	Weight%	Atomic %
B K	16.13	37.51
C K	21.01	43.98
Ti K	4.79	2.52
Cr K	0.52	0.25
Ni K	1.08	0.46
Nb L	56.46	15.28
Totals	100.00	

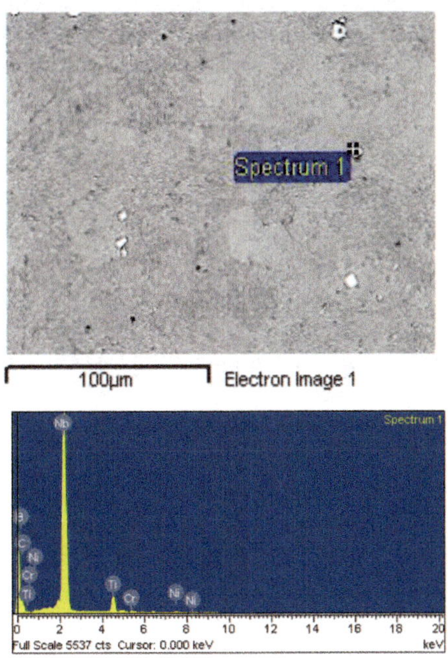

Fig. 5 EDS of NbC carbides

After solid solution heat treatment, most of the carbides were dissolved back to the γ matrix including the delta phase (δ); furthermore, it was observed that the grain size average did not significantly change neither the NbC carbides of Fig. 6a, b. During solid solution, the δ phase is dissolved; it means that mechanical properties such as yield stress, tensile stress, and hardness decrease; but elongation and toughness

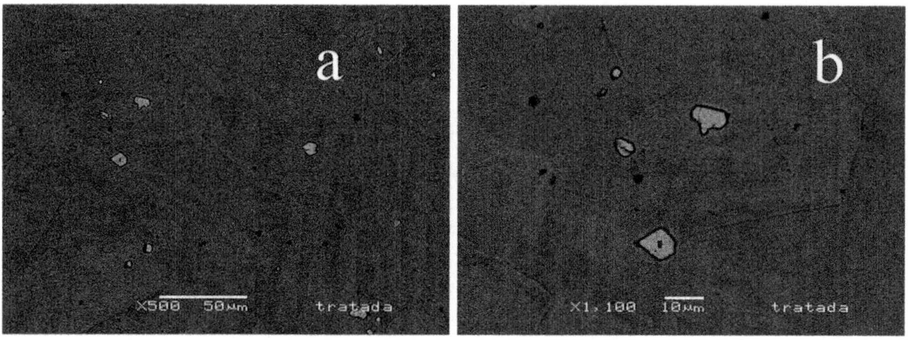

Fig. 6 SEM images of Inconel 718 after solid solution heat treatment, magnification (**a**) 500× and (**b**) 1100×

increase due to a lower amount of δ phase and bigger grains. The decrease of the last mechanical properties facilitates the machining of aeronautical pieces. A grain size average in this heat treatment was about 80 µm, and the NbC carbides conserve the same size before solid solution heat treatment. Another kind of precipitates is also observed such as NbB type by EDS analysis according to the peaks in Fig. 7.

At the same time, a modal analysis was carried out using a finite element program. This analysis is used to determine the natural oscillation frequency of the I-718, where mechanical properties are shown in Table 2. These mechanical properties were introduced together with the geometry and dimensions (mm) to the numerical modal analysis as shown in Fig. 8, in order to obtain the natural frequency of vibration. The material of the specimen is assumed with the following conditions: homogeneous, isotropic, and linear. Modal analysis result is shown in Fig. 7, where the obtained natural oscillation frequency was 20,101 Hz, near to the oscillation frequency of 20,000 Hz, which is the frequency of the ultrasonic fatigue machine.

The machining process was concerned with the following stages: facing, drilling, hammering (Aries lathe model SLB-1340), and machining (CNM lathe model WM360 brand Dynamach) (Fig. 9a, b).

After aging treatment, the sample has a typical austenitic microstructure (Fig. 10) by optical microscope with the formation of a twin as shown in Fig. 11 using a Jeol SEM model 6510LV. The grain size average increased again with the formation of twins, and there are no residues of δ phase into and on the boundaries of grains due to the solid solution treatment; the twins are responsible for the hardness, because hardening mechanism of twins reduces the grain size, blocks dislocation movements, and limits the movement of grain boundaries. The last effects contribute to the hardening of the material; the δ phase is formed at temperatures in a rate of 760–980 °C and kept stable at temperature approaching room temperature; but in a solid solution treatment above 980 °C, it has the consequence in dissolution of Nb in δ phase into the matrix. The average grain size after aging treatment is 105 µm, and mechanical properties after aging heat treatment due to precipitations of γ''

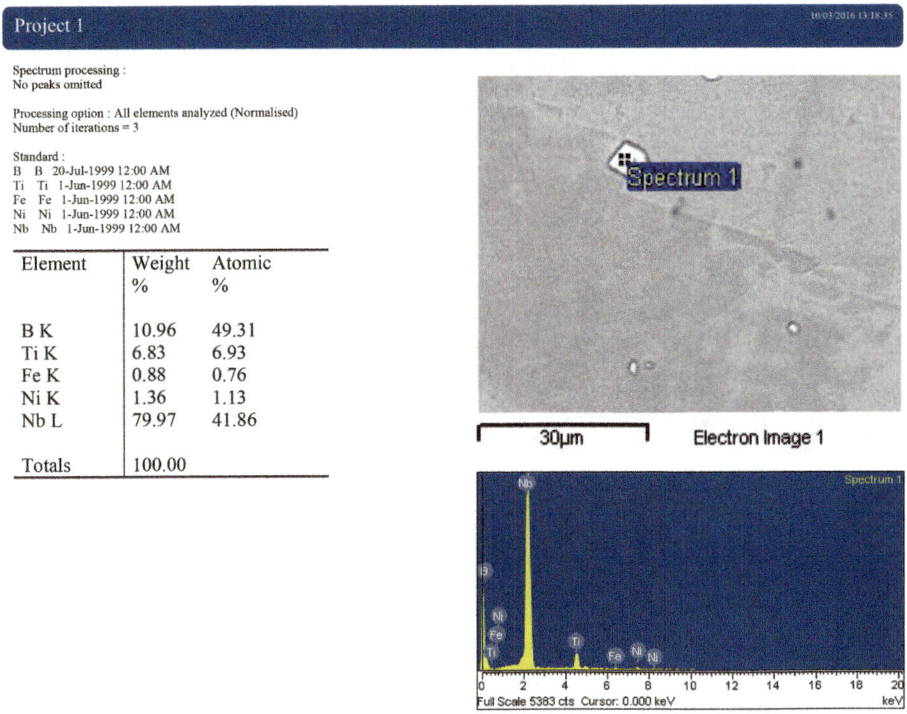

Element	Weight %	Atomic %
B K	10.96	49.31
Ti K	6.83	6.93
Fe K	0.88	0.76
Ni K	1.36	1.13
Nb L	79.97	41.86
Totals	100.00	

Fig. 7 EDS of NbB precipitates

Table 2 Mechanical properties of I-718 used in modal analysis

Density (δ)	8150 (kg/m^3)
Modulus of elasticity (young)	210 GPa
Poisson's ratio	0.294

phase (yield stress, tensile stress, and hardness) increase in opposite of other mechanical properties such as elongation and toughness.

The figure 12 shows that the heat treatments improved the precipitate formation of Nb (Fig. 12).

Heat treatments affect the mechanical properties of I-718 in correlation with microstructure changes: during solid solution hardness decreases and after aging treatment hardness increases; furthermore, mechanical properties are in correlation with distribution of precipitates. Table 3 shows the results of hardness (HV, RA, RB, and RC), annealed was as-received, after that solid solution treatment decreases hardness, and finally aging treatment increases hardness.

Microhardness tester Mitutoyo model VL 101 with a load of 1 kg by 15 s was used to carry out tests from the center of the cross section to the extremes.

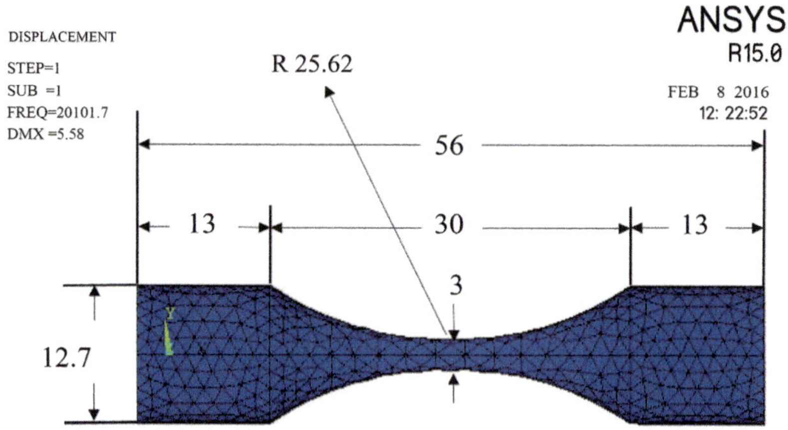

Fig. 8 Modal analysis and dimensions (mm) of I-718.

Fig. 9 (**a**) Equipment used for facing, drilling, and hammering, (**b**) CNC equipment used for machining

Fig. 10 Grain morphology after double aging treatment by optical microscope

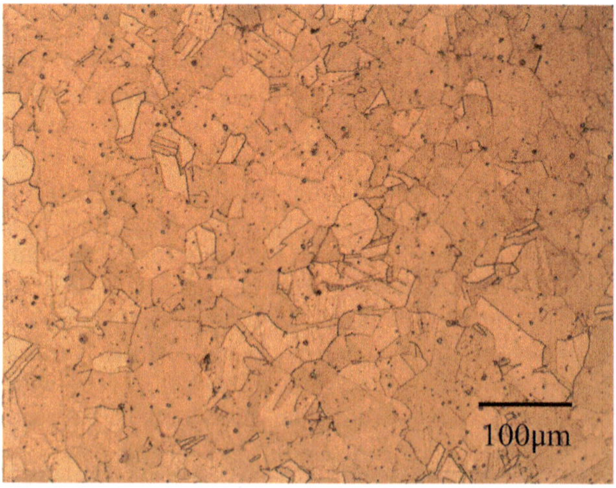

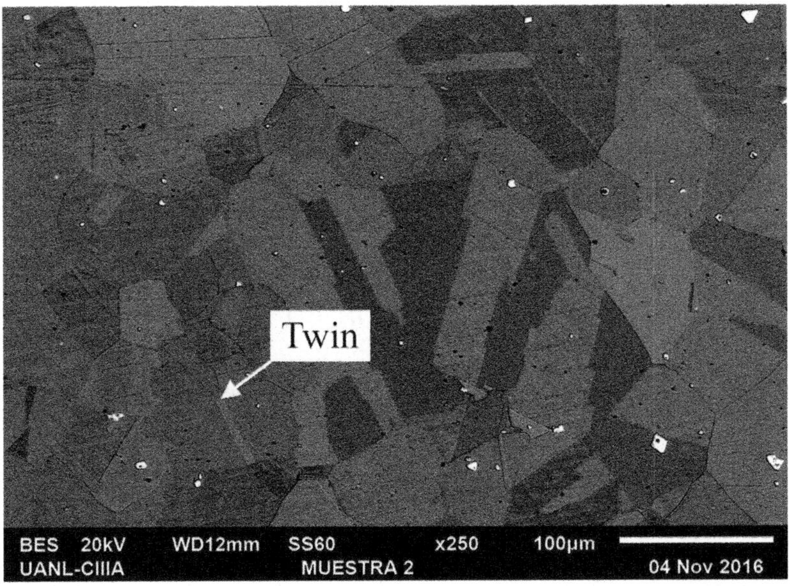

Fig. 11 After double aging treatment, it forms twins

4 Preliminary Ultrasonic Fatigue Test

A preliminary test was performed on the ultrasonic fatigue machine to corroborate that the specimen is in resonance with the ultrasonic fatigue machine. Figure 13 shows a thermographic image, where it is verified that the specimen is under the condition of resonance with the fatigue ultrasonic machine: the maximum temperature registered is localized in the neck of the specimen, where the maximum temperature and stress is located, revealing a stationary elastic wave along the testing specimen.

5 Conclusions

The following conclusions can be drawn from the present work:

1. The solid solution heat treatment dissolved the δ phase into the matrix, which was at the grain boundary as well as within the boundary grains; this process also reduced hardness to facilitate machining. The solid solution was carried out at 1078 °C.
2. By reducing the hardness of the material, it was possible to machine the specimen in the dimensions obtained through a modal analysis with the software of finite element method.

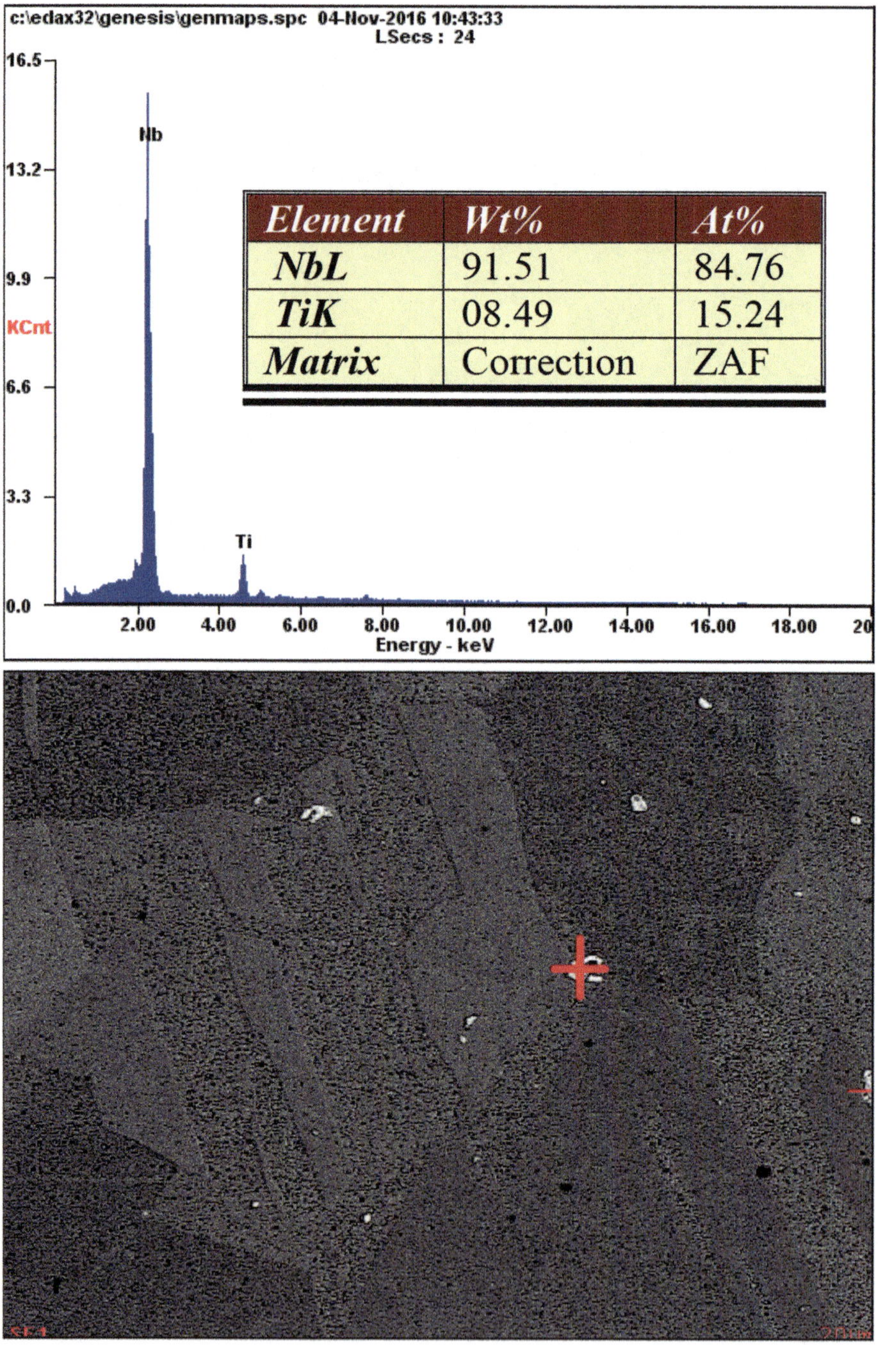

Fig. 12 EDS of Nb precipitates

Table 3 Results of hardness at different scales

Inconel 718	Hardness V	Hardness RA	Hardness RB	Hardness R C
Annealed	263	64	100	22
Solid solution	214	54	88	10
Aged	428	70	112	38

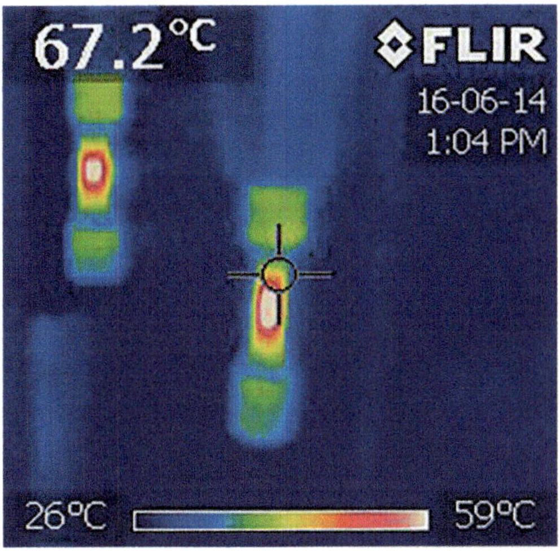

Fig. 13 Thermographic image of preliminary test of I-718 specimen in resonance with ultrasonic fatigue machine

3. The aging heat treatment increases the hardness, according to the precipitation hardening mechanism, and the hardness results are in accordance with AMS 5663 standard.
4. Thermographic images show that the maximum temperature is in the narrow section of the specimen which is coincident with the maximum applied stress. This physical evidence confirms that the specimen is in resonance with the ultrasonic fatigue machine, for future ultrasonic fatigue testing.

References

1. Mazur, Z., Luna-Ramírez, A., Islas, J. A., & Campos-Amezcua, A. (2005). Failure analysis of a gas turbine blade made of Inconel 738LC. *Engineering Failure Analysis, 12*, 474–486. doi:10.1016/j.engfailanal.2004.10.002.
2. Kargarnejad, S., & Djavanroodi, F. (2012). Failure assessment of Nimonic 80A gas turbine blade. *Engineering Failure Analysis, 26*, 211–219. http://dx.doi.org/10.1016/j.engfailanal.2012.05.028.
3. Yu-jiang, X., Mao-cai, W., Ge, Z., & Min, C. (2006). Analysis of superalloy turbine blade tip cracking during service. *Engineering Failure Analysis and Prevention, 13*, 1429–1436. doi.org/10.1016/j.engfailanal.2005.07.022.
4. Kosieniak, E., Biesiada, K., Kaczorowsky, J., & Innocenti, M. (2012). Corrosion failure in gas turbine hot components. *Journal of Failure Analysis and Prevention, 12*, 330–337. doi:10.1007/s11668-012-9571-3.

5. Blachnio, J. (2011). Analysis of causes of decohesion of a gas turbine blade made of EI867-WD alloy. *Aircraft Engineering and Aerospace Technology, 83*, 14–20. doi:10.1108/00022661111119874.
6. Balitskii, A., Vytvitskii, V., Ivaskevich, L., & Eliasz, J. (2012). The high-and low-cycle fatigue behavior of Ni-contain steels and Ni-alloys in high pressure hydrogen. *International Journal of Fatigue, 39*, 32–37. doi:10.1016/j.ijfatigue.2011.05.017.
7. Lortrakul, P., Trice, R. W., Trumble, K. P., & Dayananda, M. A. (2014). Investigation of the mechanisms of type-II hot corrosion of superalloy CMSX-4. *Corrosion Science, 80*, 408–415. doi:10.1016/j.corsci.2013.11.048.
8. Mahobia, G. S., Paulose, N., & Singh, V. (2013). Hot corrosion behavior of superalloy IN718 at 550 and 650° C. *Journal of Materials Engineering and Performance, 22*, 2418–2435. doi:10.1007/s11665-014-1278-z.
9. Loria, E. A. (1988). The status and prospects of alloy 718. *JOM, 40*, 36–41. doi:10.1007/BF03258149.
10. Zhang, P., Zhu, Q., Hu, C., Wang, C. J., Chen, G., & Qin, H. (2015). Cyclic deformation behavior of a nickel-base superalloy under fatigue loading. *Materials and Design, 69*, 12–21. doi:10.1016/j.matdes.2014.12.047.
11. Sims, C. T., Stoloff, N. S., & Hagel, W. C. (1987). *Superalloys II*. New York, NY: Wiley.
12. NACE Standard MR0175-2003. (2003). *Standard materials requirements: Metal for Sulfide stress cracking and stress corrosion cracking resistance in sour oilfield environments*. Houston: NACE International.
13. Preuss, M., Withers, P. J., & Baxter, G. J. (2006). A comparison of inertia friction welds in three nickel base superalloys. *Materials Science and Engineering A, 437*, 38–45. doi:10.1016/j.msea.2006.04.058.
14. Reicman, S., Duhl, D. N., Maurer, G., Antolovich, C., & Lund, C. (1988). Proceedings of the sixth international symposium on superalloys (pp. 3–12). TMS, Champion, PA/Warrendale, PA.
15. Loria, E. A. (1989). Proceedings of the international symposium on the metallurgy and applications of superalloys 718 (pp. 135–154). TMS, Pittsburgh, PA/Warrendale, PA.
16. Cesnik, D., Bratus, V., Kosec, B., Rozman, J., & Bizjak, M. (2008). Heat treatment and fineblankin Inconel 718, RMZ. *Materials and Geoenvironment, 55*, 163–172.
17. Yeh, A. C., Lu, K. W., Kuo, C. M., Bor, H. Y., & Wei, C. N. (2011). Effect of serrated grain boundaries on the creep property of Inconel 718 superalloy. *Materials Science and Engineering A, 530*, 525–529. doi:10.1016/j.msea.2011.10.014.
18. Kuo, C. M., Yang, Y. T., Bor, H. Y., Wei, C. N., & Tai, C. C. (2009). Aging effects on the microstructure and creep behavior of Inconel 718 superalloy. *Materials Science and Engineering A, 510-511*, 289–294. doi:10.1016/j.msea.2008.04.097.
19. American Society for Metals. (1991). *ASM handbook, Heat treating* (Vol. 4). Materials Park, OH: ASM International. ISBN 0-87170-379-3.
20. American Society for Metals. (2004). *ASM handbook, Metallography and microstructures* (Vol. 9). Materials Park, OH: ASM International. ISBN: 0-87170-706-3.

Geometry Tool Influence on the Microstructure of MMC Al-SiC Friction Stir Welds

A.F. Miranda Pérez, E. Hurtado Delgado, G.Y. Pérez Medina, F.J. García Vazquez, and P. Zambrano-Robledo

Abstract Welding as a manufacturing process on aeronautical industry is necessary for component integration. In the aircraft manufacturing, the use of aluminum alloys is widely extended. However, in these alloys if fusion welding is applied, some microstructural alterations, deleterious phases, and distortion are presented and decrease the mechanical properties of the joint. The aluminum matrix composites (AMCs) present many advantages of traditional aluminum alloys; superior properties are achieved when ceramic particles are added making them suitable for aeronautic industry. The disadvantages for AMCs with fusion welding processes are: agglomeration of the particles, lack of fusion, porosity, and many other welding defects. Recently, the solid-state welding processes are gaining acceptance due to high-quality joints, defect-free, and environmental cleaner processes, mainly the friction stir welding (FSW). Hence, FSW is commonly employed for joining these materials, and many studies have been published where the tool is the principal subject; they are mainly focused in tool wear rather than geometry influence on microstructure. Therefore, this investigation was performed in order to observe the effects of geometry differences; the aim was to examine the microstructural evolution of the welded joints. Optical and scanning microscopes (OM and SEM) were employed for the evaluation, resulting in a partial grain refinement in the thermomechanical affected zone, for the aluminum matrix and eutectic silicon, and for the fracture of the silicon carbides. In addition, orientation imaging microscopy (OIM) using electron backscatter diffraction (EBSD) technique was employed to analyze

A.F. Miranda Pérez (✉) • E. Hurtado Delgado • G.Y. Pérez Medina
Corporación Mexicana de Investigación en Materiales, Saltillo, Mexico
e-mail: argelia.miranda@comimsa.com; eduardohd@comimsa.com; gladysperez@comimsa.com

F.J. García Vazquez
Facultad de Ingeniería, UAdeC, Arteaga, Mexico
e-mail: gv.felipe@gmail.com

P. Zambrano-Robledo
Universidad Autónoma de Nuevo León, Facultad de Ingeniería Mecánica y Eléctrica, San Nicolás de los Garza, NL, Mexico
e-mail: patricia.zambranor@uanl.edu.mx

P. Zambrano-Robledo et al. (eds.), *Proceedings of the Symposium of Aeronautical and Aerospace Processes, Materials and Industrial Applications*, DOI 10.1007/978-3-319-65611-3_2

the dynamic recrystallization. Some welding defects were presented along the joint for one tool. Finally, hardness Vickers evaluation on the joints was completed resulting in differences between the two employed tools.

Keywords FSW • MMC • EBSD • Dynamic recrystallization • SEM

1 Introduction

Metal matrix composites (MMC) are drawing attention by highly demand industries including aeronautical, aerospace, military, and automotive [1] basically due to the excellent performance in aggressive environments. MMC have been developed due to the lightweight material requirement; their characteristics are based in the increment on certain properties, such as high fracture toughness, elasticity and strength, stiffness, good thermal conductivity above other composites, and excellent wear resistance [2]; this is obtained from the employment of continuous or discontinuous ceramic fibers or particles that strengthen the base metal. MMC A359/SiC/20p is a material that, besides the properties previously described, is low cost, has ease of processing, and is accessible. In addition, the enhanced properties of A359/SiC/20p compared to conventional aluminum alloy are mainly an increment of the tensile strength, which is more than 100 MPa, besides higher elasticity and ductility [3].

The expandable employment of Al matrix composites (AMCs) in aeronautical industry is mostly due to their particular properties, such as lower weight, dimensional stability, excellent wear resistance, and stiffness [4–6]. For the joints, the maintenance or increment of mechanical properties is required. Hence, fusion welding processes are not recommended, basically due to formation of detrimental intermetallic phases [7]. Advanced fusion welding processes have been employed in the AMC's joining; Bassani et al. performed laser composites joints; the proposal was intended to be a promising manufacturing process for composites due to the high beam quality, productivity, and flexibility. However, besides some recommendation and conclusions of suitable metallurgical welding quality, the welded joints present higher concentration of Al_4C_3 phase which is fragile and decrease the mechanical properties of the joint [8].

Considering the disadvantages discussed before, solid-state joining process was introduced as an alternative of joining challenging materials. Friction stir welding is a solid-state welding process invented by The Welding Institute (TWI), in the first place, to join aluminum alloys [9]; however, it is also employed to join titanium, steels, and metal composites. The process employs a cylindrical tool which is frictionally rotating producing heat between the parent metal and the tool, which is moved along the seam weld. The friction generates plastic deformation at low temperature not exceeding the melting point of the base material [10]. Dynamic recrystallization due to the severe plastic deformation produces grain refinement as a resulting microstructure. The tool is determinant in this process such that any geometry modification can be detrimental or beneficial for the final microstructure and hence the mechanical properties. Krasnowski et al. [10] studied the influence of varying the tool shape in aluminum alloys, in which after evaluating three different probe shapes, concluded that the shape severely influences the mechanical properties in the Al 6082-T6. Other investigators also mention grain size modifications due to tool shoulder [11], but most of the work done is on aluminum alloys.

This research was developed varying the pin geometry of the FSW tool by heat-treating a conventional D2 tool steel until it achieves enough hardness for metal forging. Therefore, this investigation was developed considering that the main modifications on the microstructure due to geometry tool variation in the AMC are not broadly explored.

2 Experimental Procedure

Plates of AMC A359 with 20% SiC particles of reinforcement (A359/SiC/20p) and 5 mm of thickness were employed; chemical composition is shown in Table 1. This material was obtained by the company MC 21, and the mechanical properties are shown in Table 2. The samples were prepared with a rectangular shape 88 mm × 30.5 mm; for friction stir welding, a computerized numerical control machine was employed (Brand HAAS, model VM2) (Fig. 1). The tools for the joining were elaborated at COMIMSA with geometry variations, the first one is based in the commercial tool Trivex [12] developed by the TWI, and the second has a slight modification in the pin. The diagonal curved notches in the second tool were suggested in order to act as a reservoir of plastic base metal and behave as a pin wear protection. Both of the tools consist of a concave cylindrical shoulder of 75 mm of height and 25.4 mm of diameter, and the second tool has a diagonal curves incision in the pin of 5.5 mm of height (Fig. 2).

Table 1 Chemical composition (wt. %) of the AMC A359/SiC/20p

Chemical composition AMC A359/SiC/20p							
Si	Fe	Cu	Mn	Mg	Zn	Ti	Al
8.5–9.5	0.2	0.2	0.1	0.50–0.70	0.1	0.2	Balance

Table 2 AMC A359/SiC/20p mechanical properties [3]

Mechanical properties							
Density (g/cm^3)	Yield strength (MPa)	Ultimate tensile strength (MPa)	Elongation %	Young modulus (GPa)	Thermal conductivity (W/mK)	Hardness (HV)	Fracture toughness (MPa)
2.72	214	262	1.9	98.6	150	73.94 HV	15.9

Fig. 1 CNC machining center

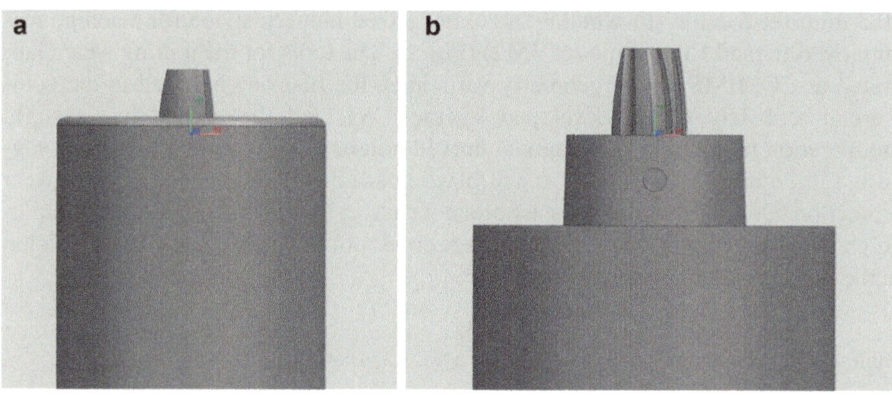

Fig. 2 Tools fabricated in COMIMSA, (**a**) triangular tool and (**b**) diagonal curved notch tool

The FSW tools were manufactured in D2 tool steel, in order to increase the hardness and wear of the tools, and were subjected into an isothermal heat treatment, described in (Fig. 3), achieving a final hardness of 54 Rockwell C. D2 tool steel was selected since it has high surface wear resistance with a sufficiently tough core, after a suitable heat treatment, and can reach high hardness for metalworking [13].

The FSW parameters were proposed according to the investigation developed by T. Prater et al. [14] and those performed by O. Cuevas Mata et al. [15] and are shown in Table 3. In order to differentiate between the two tools, they were named as original design (O) and notched (N). Moreover, a set of experiments was performed; for

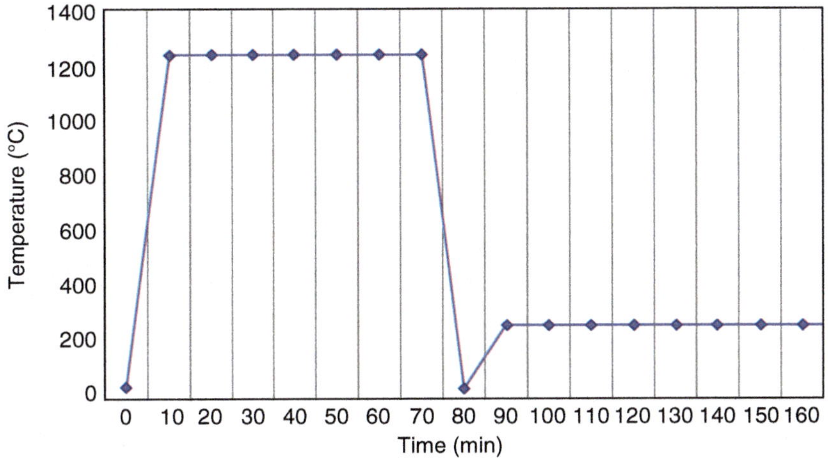

Fig. 3 Heat treatment ranges for both tools

Table 3 Welding parameters used for the joint of the material by the FSW process

Sample	1a	1b	2	3	4	5	6	7	8
Tool	O	O	O	O	N	N	N	O	O
Welding speed (mm/m)	75	75	50	50	75	50	40	40	50
Rotational speed (rpm)	1000	1250	1000	1400	1000	1400	1600	1600	1500

the microstructural analysis, the samples 3 and 5 were selected. Cross sections of the samples were cut for metallographic examination. The microstructural characterization consisted of grinding and polishing using Weck as reactive etchant for 20s.

The samples were examined in an Olympus Optical Microscope and scanning electron TESCAN MIRA microscope using a Bruker EDAX detector for the mapping scanning analysis. For the electron backscatter diffraction (EBSD), a special cross section ion polisher was employed for 8 h, and the EBSD examination was performed using the TSL-OIMA software. Lastly, the microhardness test of the samples was carried out in Vickers with 24 indentations of 1 mm distance of each sample with a diamond-shaped nanoindenter brand Tukon 2500-5 Wilson Hardness 50x/0.55.

3 Results and Discussion

The base metal consists of an aluminum matrix with silicon carbide particles approximately of 20 μm thick; some eutectic silicon is evident as well as in the micrographs presented in Fig. 4. The silicon carbide particles are distributed suitably concentrated in the interdendritic regions.

Fig. 4 Secondary electron micrograph of the AMC

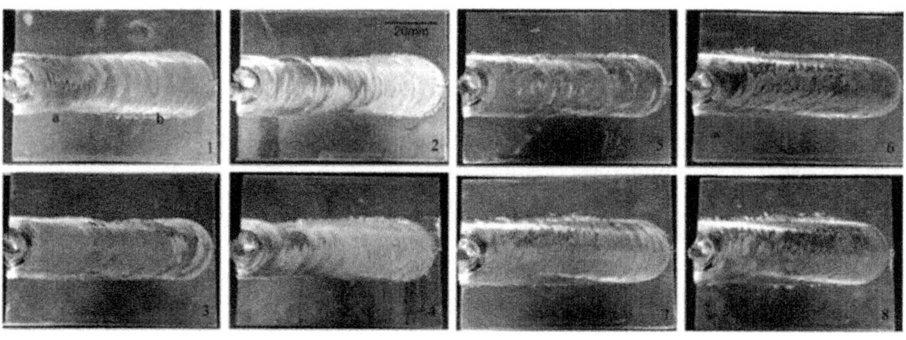

Fig. 5 Composite A359/SiC/20p joined by the different process parameters

The welded samples are presented in Fig. 5, some of them evidence surface defects as flashing or initial crack which on visual inspection can conduct a wormhole [16]. The wormhole is an internal defect which is formed by an insufficient stirring between the materials and excessive heat input. In the Table 3, the samples highlighted were the ones which present higher welding quality since no surface discontinuity or welding defect was evident.

The result of the SEM/EDX analysis for the characterized surface of the sample is shown in Fig. 6. The thermomechanical affected zone (TMAZ) was selected for the analysis. Elemental distribution of secondary phases was determined using the mapping analysis of constituent elements Al, Fe, Cu, C, Si, Ti, Mg, and Mn. Silicon

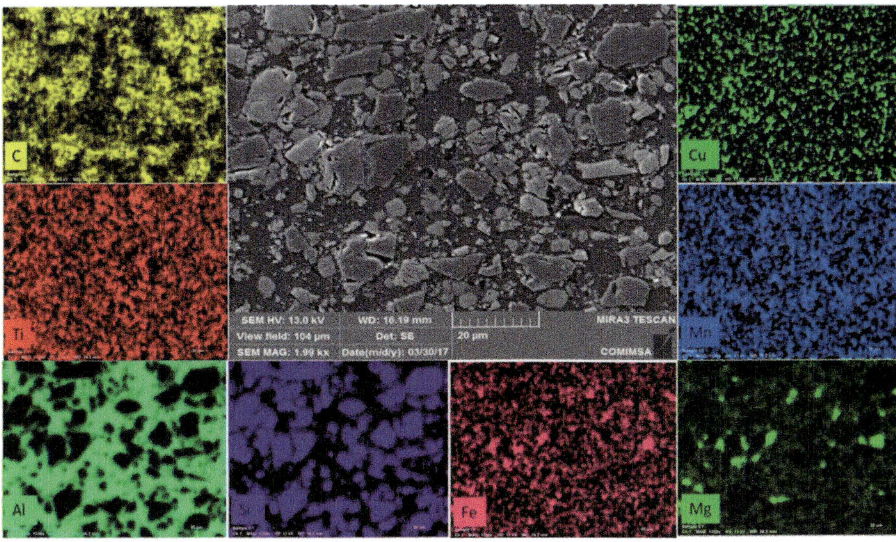

Fig. 6 Scanning electron micrograph with EDX mapping of the FSW A359/SiC/20p

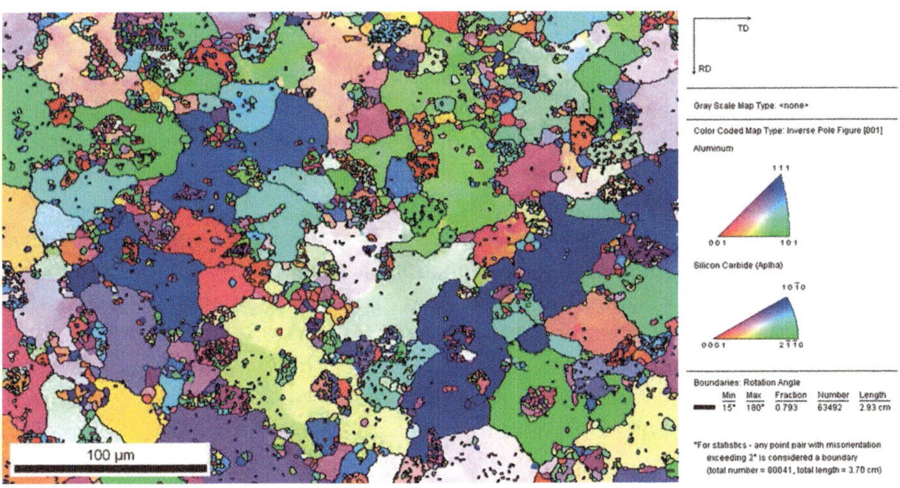

Fig. 7 EBSD inverse pole figure evidence the TMAZ, showing the recrystallized grains

carbides are distributed throughout the matrix, and some of them were fractured as shown in Fig. 6. Amounts of Fe and Mg are evident in the EDX analysis located near to the SiC. β-phase was not identified in the EDX analysis; this secondary phase is a deleterious Al_3Mg_2 compound which can decrease the mechanical properties of the joint [17].

Figure 7 presents the inverse pole figure of the thermomechanical affected zone. This technique is an effective tool for grain evolution analysis, texture, and recrystallization and also for the study of materials subjected into thermomechanical processing [18].

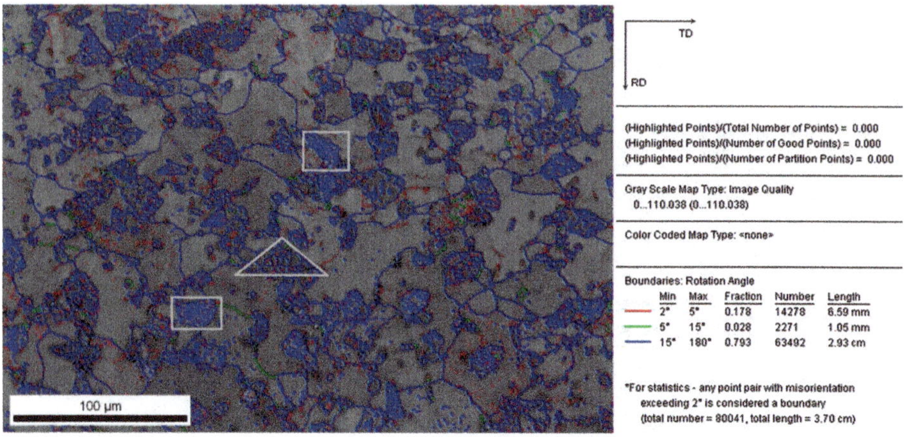

Fig. 8 EBSD boundary map shows the recrystallized grains

The EBSD image presents the grain structure modifications due to FSW process. The grains are considerably finer in comparison with the base metal, partitioning and fracture of the SiC are evident, and the reinforcement particle fracture provides a positive effect since it acts as new grain nucleation sites [15].

The formation of fine new grains is the result of dynamic recrystallization due to intense plastic deformation. This process produces flow of material which promotes the grain structure to be reoriented due to the applied stress; these grains split into regular arrays or nearly parallel bands. In Fig. 8, recrystallized aluminum grains were indicated in rectangular cases; the fracture particles of SiC are enclosed in a triangle instead.

From Fig. 8, it is evident that most of the new grains have a misoriented angle greater than 15°, meaning that their orientations are completely different from the parent ones. Some new grain boundaries, with a misoriented angle less than 15°, were forming in the larger grains, where the deformation was not enough to produce a complete recrystallization, because the analyzed region is a transition zone between the unaffected material and the TMAZ. The grain size of the recrystallized grains is very small (around 1 µ) as compared with the unrecrystallized grains. The fraction of recrystallized grains in this region is close to 20%; this fraction is increased up to 100% in the stirring zone. In this figure, some grains are completely recrystallized, forming clusters of small grains, but some others are partially recrystallized, showing small grains near the original grain boundaries or isolated tiny grains in the interior of large grains. The cracked SiC particles (the darker grains in the figure) have produced large misoriented-angle boundaries, which indicate that the strain of the matrix forced the fracture of the ceramic particles, unable to be plastically deformed, but the internal stress compacted them, holding their fragments together, apparently as a unique particle, but the EBSD has revealed their fragmented nature.

Microhardness evaluation was performed in the selected samples and is shown in Fig. 9. The stirred zone reaches the higher hardness values due to the agglomeration of the SiC particles that were fractured, and the refined grains, near the TMAZ some grains, were recrystallized with finer grains due to the temperature reached and the plastic deformation.

Pin tool wear is presented in Fig. 10; Trivex tool was originally with three convex sides which were intended to avoid material waste. The convex geometry was destroyed, and cylindrical shape was obtained after the welding process. In the case of the diagonal curved notched tool which is presented in Fig. 10b, the tool was protected and filled by the base metal after the plastic deformation is evident that wear of the second tool is minor. Volume loss was approximately 30% more in the original tool than in the diagonal curved notched.

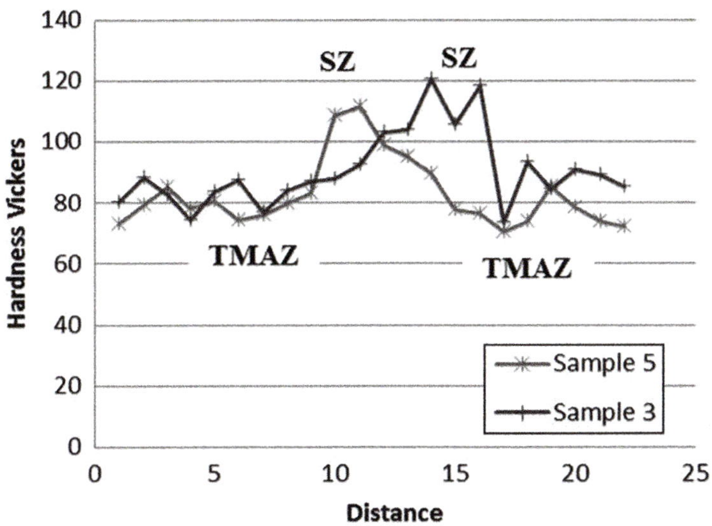

Fig. 9 Vickers microhardness profile of sample 3 and 5

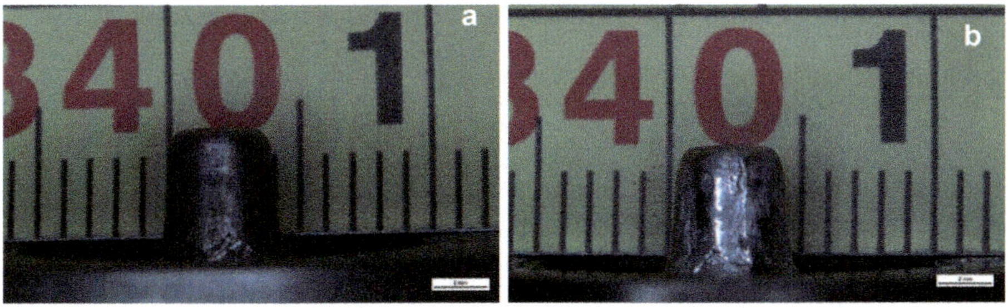

Fig. 10 Tool after welding (**a**) original design and (**b**) notched geometry

4 Conclusions

The tool geometry influence on the microstructure in an AMC A359/SiC/20p was investigated. The pin modification effect in the friction stir welding application was discussed. Based on the presented study, the important results can be summarized as follows:

Some internal defects were observed due to the variation of parameters and the tool employed. The principal one was wormholes. Surface defects as flashing was presented for both tools when higher welding speed is reached.

Silicon carbide particles were fractured during the welding process leading to some accumulation of Mg and Fe around these particles.

EBDS IPF evidences the dynamic recrystallization due to intense plastic deformation; the fractured ceramic particles serve as a preference nucleation site.

The new grains are clearly differenced from the parent ones as shown in the EBSD image.

Microstructural changes due to solid-state welding process improved the microhardness of the joint; besides the obtained recrystallization, this can be also due to the absence of β-phase in the aluminum matrix. The sample with better results was the original one.

The second tool suffers less wear than the original one and is approximately 30% of volume loss.

References

1. Tian, W., Qi, L., Su, C., Zhou, J., & Jing, Z. (2016). Numerical simulation on elastic properties of short-fiber-reinforced metal matrix composites: Effect of fiber orientation. *Composite Structures, 152*, 408–417. doi:10.1016/j.compstruct.2016.05.046.
2. Nishida, Y. (2013). *Introduction to metal matrix composites*. Tokio: Springer. doi:10.1007/978-4-431-54237-7.
3. Kothari, M. A. (2005). *Welding of cast A359/SiC/10p metal matrix composites* (p. 97). College Station: Texas A&M University. Master thesis.
4. Alam, S. N., & Kumar, L. (2016). Mechanical properties of aluminium based metal matrix composites reinforced with graphite nanoplatelets. *Materials Science and Engineering A, 667*, 16–32. doi:10.1016/j.msea.2016.04.054 5.
5. Singh, R., & Singh, G. (2015). Investigations of al–SiC AMC prepared by vacuum moulding assisted stir casting. *Journal of Manufacturing Processes, 19*, 142–147. doi:10.1016/j.jmapro.2015.06.011.
6. Chen, X.-H., & Yan, H. (2016). Solid–liquid interface dynamics during solidification of al 7075–Al2O3np based metal matrix composites. *Materials and Design, 94*, 148–158. doi:10.1016/j.matdes.2016.01.042 7.
7. Chen, M., Wu, C., & Gao, J. (2002). Welding of SiC particle reinforced 6061 Al matrix composite with pulsed TIG. *Transactions of Nonferrous Metals Society of China, 12*, 805–810.
8. Bassani, P., Capello, E., Colombo, D., Previtali, B., & Vedani, M. (2007). Effect of process parameters on bead properties of A359/SiC MMCs welded by laser. *Composites Part A: Applied Science and Manufacturing, 38*, 1089–1098. doi:10.1016/j.compositesa.2006.04.014.
9. Thomas, W. M. (1991). 9125978 GB Patent.

10. Krasnowski, K., Hamilton, C., & Dymek, S. (2015). Influence of the tool shape and weld configuration on microstructure and mechanical properties of the al 6082 alloy FSW joints. *Archives of Civil and Mechanical Engineering, 15*, 133–141. doi:10.1016/j.acme.2014.02.001.
11. Casalino, G., Campanelli, S., & Mortello, M. (2014). Influence of shoulder geometry and coating of the tool on the friction stir welding of aluminium alloy plates. *Procedia Engineering, 69*, 1541–1548. doi:10.1016/j.proeng.2014.03.153.
12. Colegrove, P. A., & Shercliff, H. R. (2004). Development of Trivex friction stir welding tool part 1–two-dimensional flow modelling and experimental validation. *Science and Technology of Welding and Joining, 9*, 345–351. doi:10.1179/136217104225021670.
13. Conci, M. D., Bozzi, A. C., & Franco, A. R. (2014). Effect of plasma nitriding potential on tribological behaviour of AISI D2 cold-worked tool steel. *Wear, 317*, 188–193. doi:10.1016/j.wear.2014.05.012.
14. Prater, T. (2014). Friction stir welding of metal matrix composites for use in aerospace structures. *Acta Astronautica, 93*, 366–373. doi:10.1016/j.actaastro.2013.07.023.
15. Cuevas Mata, O., Miranda Pérez, A. F., García Vázquez, F. J., Pérez Medina, G. Y., & Reyes Valdés, F. A. (2015). Effect of FSW parameters on microstructure of Aluminum matrix composites joints. In *Materials characterization* (pp. 139–146). Cham: Springer. doi:10.1007/978-3-319-15204-2_14.
16. Mishra, R. S., & Mahoney, M. W. (2007). *Friction stir welding and processing* (p. 352). Materials Park, OH: ASM International.
17. Choi, D.-H., Ahn, B.-W., Quesnel, D. J., & Jung, S.-B. (2013). Behavior of β phase (Al3Mg2) in AA 5083 during friction stir welding. *Intermetallics, 35*, 120–127. doi:10.1016/j.intermet.2012.12.004.
18. Medina, G. P., Ferreira, H. L., Robledo, P. Z., Pérez, A. M., & Valdés, F. A. R. (2016). Microstructural development in a TRIP-780 steel joined by friction stir welding (FSW): Quantitative evaluations and comparisons with EBSD predictions. *Solda Inspeç, 21*, 146–155. doi:10.1590/0104-9224/si2102.04.

Study of Decomposition of Martensite in FV535 Steel

L. Guerra-Fuentes, M.A.L. Hernandez-Rodriguez, P. Zambrano-Robledo, A. Salinas-Rodriguez, and E. Garcia-Sanchez

Abstract The decomposition of martensite and precipitation presented during this process in the FV535 high-chromium martensitic steel and the effect on mechanical properties were studied. FV535 steel is used in petrochemical plants, gas turbine engineering, aircraft, and aerospace industries. The decomposition of martensite has been divided in six different stages, occurring at different temperature ranges, which can vary depending on the steel composition and microstructure; these stages were analyzed by DSC, and in order to analyze the change in mechanical properties, the FV535 steel was characterized by microhardness tests. The precipitation phenomena that occurred in this type of steel were analyzed by DSC and transmission electron microscopy and were correlated with the change in mechanical properties at high temperatures.

Keywords Martensitic steel • Tempering • TEM

1 Introduction

The material analyzed in this research is the Firth-Vickers FV535, which is a 12Cr-Mo-VNbWCo steel with application in aeronautical components with excellent mechanical properties at relatively high temperatures [1, 2]. This steel shows a martensitic structure, and when it is necessary to modify its mechanical properties, a tempering process is performed. The stages of the tempering are:

(i) Consists of the segregation of carbon atoms.

L. Guerra-Fuentes (✉) • M.A.L. Hernandez-Rodriguez
P. Zambrano-Robledo • E. Garcia-Sanchez
Universidad Autónoma de Nuevo León, Facultad de Ingeniería Mecánica y Eléctrica, San Nicolás de los Garza, NL, Mexico
e-mail: angela.liz@hotmail.com

A. Salinas-Rodriguez
CINVESTAV Unidad Saltillo, Ramos Arizpe, Coah, Mexico

P. Zambrano-Robledo et al. (eds.), *Proceedings of the Symposium of Aeronautical and Aerospace Processes, Materials and Industrial Applications*, DOI 10.1007/978-3-319-65611-3_3

(ii) Refers to the precipitation processes that occur below 80 °C and consist of the formation of clusters of carbon atoms in the iron matrix.
(iii) Between 80 and 180 °C, the precipitation of ε-carbide occurs.
(iv) Takes place in the temperature range of 200 and 350 °C and is associated with the transformation of retained austenite into ferrite and cementite.
(v) Between 250 and 500 °C, involves the conversion of the transition carbide into cementite.
(vi) Takes place in temperature range of 450 and 700 °C and is associated with the dissolution of cementite and the formation of more stable precipitates and takes place at temperatures between 450 and 700 °C [3, 4].

All of these stages promote microstructural modifications in FV535 steel which were analyzed by DSC, while the precipitate that occurs in the sixth stage of tempering was examined by conventional transmission electron microscopy (TEM). On the other hand, mechanical properties were analyzed through microhardness tests.

2 Experimental Procedure

The chemical composition of the steel studied is presented in Table 1. FV535 steel was obtained from a billet and was preheated at 670 °C for 30 min, after that heated up to 1170 °C for 1 h, and then quenched in oil; finally, the material was submitted to two tempering heat treatments, 610 °C for 4 h and cooled in air and after 620 °C for 4 h and cooled in the same way.

In order to analyze the different stages of martensite tempering, the material was solubilized at 1020 °C for 15 min and then quenched in air; after that, the material was analyzed through DSC using a differential scanning calorimeter NETZSCH model DSC 404C Pegasus with platinum pans; the mass for each sample was between 23 and 27 mg. DSC analyses were performed under argon atmosphere and heated up to 650 °C using heating rates of 15 and 20 °C/min.

A Titan 80–300 transmission electron microscope was used to observe the precipitates formed in the last stage of tempering.

In order to analyze the change of mechanical properties, microhardness tests were performed at different tempering temperatures using a Shimadzu tester doing 25 indentations in each sample.

3 Discussion and Results

Figure 1a shows the thermogram obtained at 650 °C when the material was heated at 15 °C/min; the precipitation starts at 446 °C and ends at 575 °C with an enthalpy change about 2.84 J/g, while in Fig. 1b, the thermogram obtained at a rate of 20 °C/

Table 1 Chemical composition of FV535 Steel/% wt

C	Si	Mn	P	S	Cr	Mo	Ni	B	Co	N	Nb	V
0.09	0.42	0.83	0.017	0.0014	10.53	0.70	0.35	0.0070	5.72	0.011	0.33	0.21

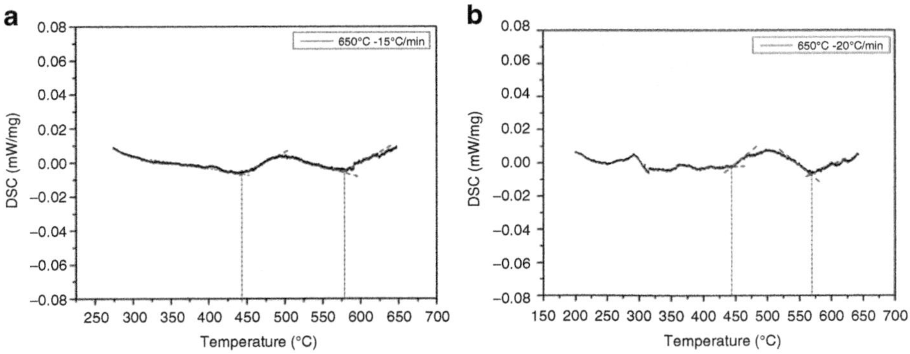

Fig. 1 (**a**) Thermogram obtained at 15 °C/min, (**b**) Thermogram obtained at 20 °C/min

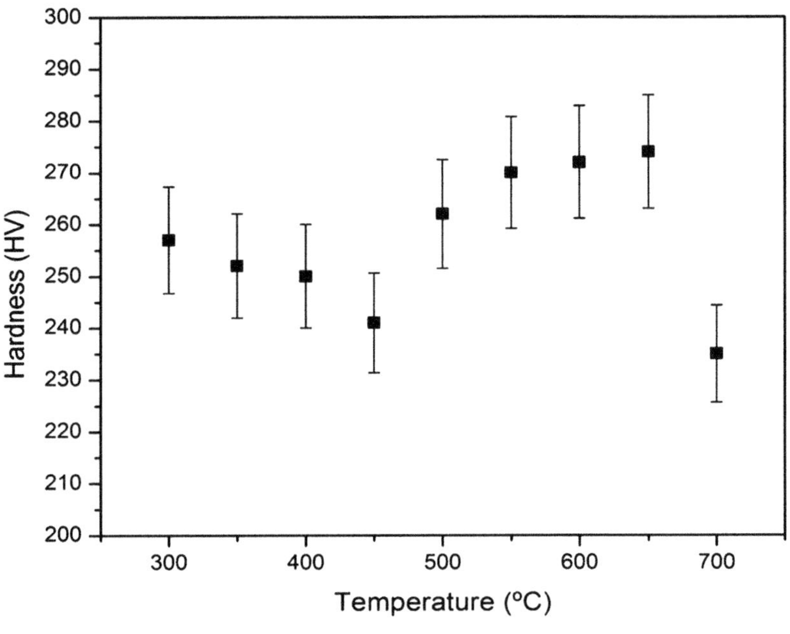

Fig. 2 Microhardness tests after each heat treatment

min can be observed; at this velocity, the precipitation occurs between 445 and 572 °C with an enthalpy change about 2.84 °C. With these results, it can be concluded that the precipitation temperature does not show a significant change when the heating rate is increased.

Figure 2 shows the results of microhardness tests after different tempering temperatures; a decrement of hardness values due to the effect of the first stages of the tempering of martensite is observed [5]. An increase in the hardness is observed at 500 °C due to the precipitation of carbides. Some alloy elements like niobium

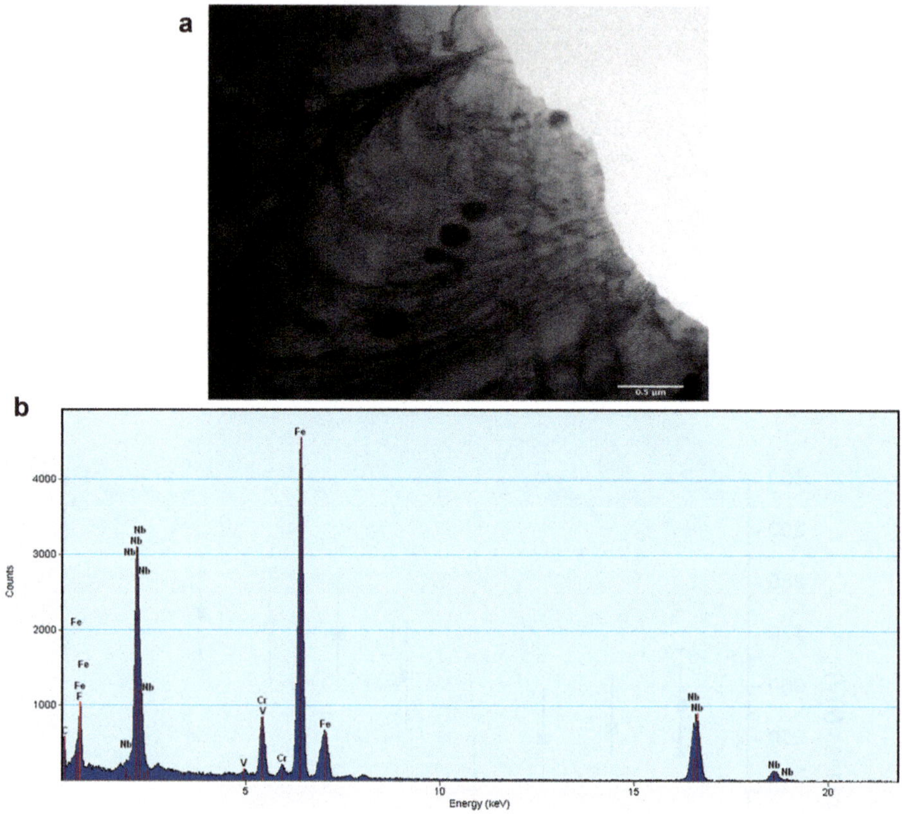

Fig. 3 (**a**) Dispersion of precipitates analyzed by TEM, (**b**) EDS microanalysis of the precipitate (**a**)

promote the formation of carbides, which were observed by TEM. Figure 3a shows the dispersion of this type of precipitates [6, 7]; Fig. 3b shows representative results of EDS analysis of the precipitates. Finally, the decrement of hardness is due to the formation of ferrite about 700 °C.

4 Conclusions

The objective of this research was to show a thermic analysis of the decomposition of martensite. Through a TEM analysis the precipitation presented in the last stage of the tempering can be observed. The mechanical response was analyzed by microhardness. The hardness of the martensite at different tempering temperatures was measured and exhibits a decrement at higher temperatures.

References

1. Klueh, R. L., & Harries, D. R. (2001). *High chromium ferritic and martensitic steels for nuclear applications* (pp. 5–12). West Conshohocken: ASM.
2. Erice, B., Gálvez, F., & Cendón, D. A. (2012). Flow and fracture behavior of FV535 steel at different triaxialities, strain rates and temperatures. *Engineering Fracture Mechanics, 79*, 1–17.
3. Speich, G. R., & Leslie, W. C. (1972). Tempering of steel. *Metallurgical Transactions, 3*, 1043–1054.
4. Van Genderen, M. J., Isac, M., Bottger, A., & Mittemeijer, E. J. (1997). Aging and tempering behavior of iron-nickel-carbon and iron-carbon martensite. *Metallurgical and Materials Transactions A, 28*, 545–561.
5. Salemi, A., & Abdollah-Zadeh, A. (2008). The effect of tempering temperature on the mechanical properties and fracture morphology of a NiCrMoV steel. *Materials Characterization, 59*, 484–487.
6. Wang, T., Cui, C., Jia, K., & Zhao, L. (2012). Effects of the nanostructured Fe-V-Nb modificators on the microstructure and mechanical properties of Si-Mn steel. *Journal of Nanomaterials, 2012*(2012), Article ID 140405, 6 pages, http://dx.doi.org/10.1155/2012/140405.
7. Bejar, L., Hernandez, C. A., Mancilla, J. E., Paraguay, F., & Espinosa, F. (2002). Estudio mediante microscopía electrónica de transmisión de precipitados de carburo de niobio en un acero microaleado. *Revista latinoamericana de metalurgia y materiales, 22*, 21–25.

On the Effect of Hysteresis in Shock Isolators for Aeronautical Applications

Diego Francisco Ledezma-Ramírez and Pablo Ernesto Tapia-González

Abstract An overview of the use of shock isolators based on dry friction is briefly presented, and the possibilities for the development of a more efficient shock isolation system are discussed. Cable isolators, also known as wire rope isolators, are studied and their characteristics are discussed. Such isolators present nonlinear stiffness behaviour in different directions, i.e. tension-compression, roll and shear, as well as dry friction damping. These isolators show hysteresis when subjected to cyclic loading and are known for being excellent shock isolators. However, little is known about the actual dynamic behaviour under shock loading. The Bouc-Wen model of hysteresis is widely used for representing the dynamic behaviour of nonlinear systems and structures under cyclic loading, as it can reproduce a variety of softening and hardening hysteresis loops by proper selection of its parameters. Recently, the interest towards the use of this model has increased, particularly for the modelling of dry friction isolators, such as wire rope springs. This chapter presents a theoretical analysis of the shock response of isolators based on the Bouc-Wen model, considering different scenarios for hardening and softening parameters. Different parameters such as absolute and relative response are studied, and the advantages and issues are discussed. It is concluded that the combination of nonlinear stiffness and damping could lead to improved vibration and shock isolation.

Keywords Shock response • Dry friction • Hysteresis • Bouc-Wen • Wire rope isolator

1 Introduction

Undesirable effects as fatigue, noise and wear are common results of transient mechanical vibration, and this is particularly important in environments such as military, naval and aerospace applications, where it is common to mitigate these effects

---

D.F. Ledezma-Ramírez (✉) • P.E. Tapia-González
Universidad Autónoma de Nuevo León, Facultad de Ingeniería Mecánica y Eléctrica,
Av. Universidad SN, San Nicolás de los Garza, NL, Mexico
e-mail: diego.ledezmard@uanl.edu.mx

P. Zambrano-Robledo et al. (eds.), *Proceedings of the Symposium of Aeronautical and Aerospace Processes, Materials and Industrial Applications*,
DOI 10.1007/978-3-319-65611-3_4

using vibration isolation. Aeronautical applications are of particular interest, since avionic components are mounted in isolators to prevent potential damage. Wire rope isolators (WRI) or cable isolators are commonly used in those applications due to their high energy storage and dissipation properties. These isolators are constructed using a series of steel strands twisted around a core strand, and the resulting wire rope is arranged in a leaf or helical fashion and presents a great capability for energy dissipation due to the friction created between the wire strands as the cable twists when the isolator is loaded and unloaded. However, most of the isolation designs based on wire ropes are empirical or based in classic isolation theory, which does not reflect the nonlinear properties typically observed in these isolators.

One of the first applications of dry friction in the isolation of shocks was conducted by Mercer [1] who designed an optimal isolator based on the principle of adjustable friction, obtaining better results than previous studies where polymers were studied, as Snowdon considered [2]. Molyneux [3] studied the behaviour of different arrangements with low stiffness springs with limited displacement ranges for mechanical vibration isolation in aeronautical applications. Eshleman et al. [4, 5] studied the response of different configurations of impact isolators as coil springs, fluids, pneumatics, ring springs and friction dampers demonstrating their high capacity in terms of energy dissipation. Tinker et al. [6] published studies on nonlinear stiffness and damping, in which an analytical model for the description of the hysteresis loop, that is common in systems with nonlinear stiffness, was derived. This study focused on the wire springs; the authors noted that the spring wire strands tend to separate when compressing, whilst when tensioned, a greater number of contact points exist and the resistance to relative motion is increased; thus, the force-displacement function behaves differently in tension than in compression. Subsequently, Cutchins et al. [7] continued their investigations in which they sought to develop a semiempirical analytical model, which could fully describe the impact isolator behaviour arranged by wire coil springs under axial loads. Other applications of the wire spring are in the civil engineering. Gerges and Vickery [8] designed and tested experimentally a tuned mass damper using wire springs. Hogea and Bausic [9] performed experimental tests on cable isolators subjected to axial excitation, obtaining data about dynamic stiffness and damping under harmonic loads.

The Bouc-Wen model of hysteresis is widely used to describe nonlinear systems that experience hysteretic behaviour, due to the ability to represent analytically a variety of hysteresis patterns, i.e. loops [10]. The parameters that represent the Bouc-Wen restoring force can be obtained from experimental data in forced or displacement-controlled experiments by system identification procedures, such as Kalman filters, differential evolution, genetic algorithms, particle swarm optimization, etc. [11]. The model has been extensively studied, for instance, the nonlinear characteristics, bifurcation and chaos were analysed by Li et al. [12] for a variety of parameters giving typical loops for hardening and softening systems, finding that chaotic behaviour is possible. The model has been widely used to represent the dynamic behaviour of wire rope or cable isolators. Ni et al. [13] developed dynamic test in wire rope isolators under shear, roll and tension-compression loads, finding symmetric softening and hardening stiffness loops for shear and roll and asymmetric stiffness during tension-compression, using a modified Bouc-Wen model to represent the dynamics and the further prediction of the response. Gerges [14] developed a mathematical model of the restoring force of a cable isolator for tension-compression, demonstrating that the force-displacement relationship is frequency independent but highly nonlinear and hysteretic in nature. The semi-analytical model agreed with the experimental data. Di Massa et al. [15] developed a prototype of wire rope isolator restraining the vertical motion with a ball transfer unit, allowing only horizontal motion. Based on experimental results, the authors identified the restoring force by means of the Bouc-Wen model, simulating the nonlinear response of the system. Wang et al. [16] analysed a cable isolator for shear, roll and tension-compression, fitting the experimental data using a modified Bouc-Wen model, finding that the system under compression presents highly nonlinear behaviour, whilst the shear and roll motion can be approximated as linear. Recent studies by Carboni et al. [17] have introduced the use of nitinol strands and steel wires, into the form of a vibration absorber, observing hardening pinching stiffness, quasilinear softening and intermediate behaviour for different configurations, identifying the parameters from experimental studies into a modified Bouc-Wen model to account for pinching, using a differential evolutionary algorithm.

Despite the recent interest on cable isolators, most of the published works are related to parameter identification and modelling, with few works related to vibration isolation properties. A recent paper by Guzmán-Nieto et al. [18] explored damping properties and briefly presented the shock response of several cable isolators. Other recent studies have analysed theoretical and experimental behaviour of nonlinear stiffness to shock response, which is similar to the nonlinear behaviour observed in cable isolators [19, 20]. Based on the previous introduction, the variety of scenarios that can be modelled after the Bouc-Wen model is particularly useful for studying the dynamics of cable isolators; however, no theoretical analysis of the shock response has been reported to the knowledge of the authors. This chapter aims to present a theoretical analysis of the shock response of nonlinear isolators modelled after the Bouc-Wen approach, considering different parameters covering softening and hardening loops. The system is excited with a base displacement half-sine pulse of different durations and amplitudes, analysing the response parameters such as maximum, residual and relative, discussing the effect of the softening and hardening loops in the response. Therefore, an insight into the shock isolation properties and the nonlinear stiffness and damping behaviour is presented, in order to propose ideas for a further mathematical model to predict the response of such isolators. The development of a mathematical model will enable future applications in isolation technology, vibration and noise control, particularly aerospace applications, i.e. landing gears, unmanned aircraft systems and avionics.

2 Background

A particular type of systems that can be approximated using the Bouc-Wen model are cable or wire rope vibration isolators. Current work at the Structural Dynamics Laboratory is being developed regarding the experimental characterisation and

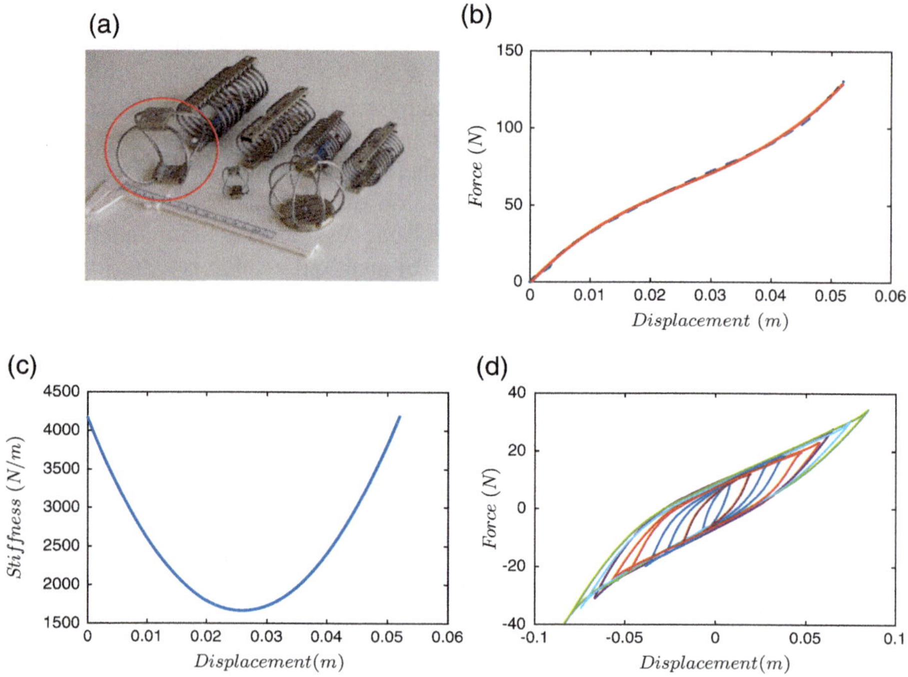

Fig. 1 Static and dynamic characteristics of wire rope isolators. (**a**) Commercial samples of WRI, the plots correspond to the isolator circled in *red*, (**b**) static force-displacement plot, (**c**) static stiffness-displacement plot, (**d**) hysteresis loops resulting from cyclic loading at a fixed frequency, with varying input amplitude

subsequent modelling of these isolators for shock loading; therefore, a first step is to study their cyclic behaviour; thus, the theoretical response of hysteretic models is important for predicting the behaviour under shock inputs. Several static and dynamic measurements have been performed in order to characterise the properties of the WRI. Figure 1a presents several examples of commercially available WRI, showing different configurations. Under static loading, the force deflection curve is nonlinear, as depicted by the broken line in Fig. 1b. The force-displacement curve can be modelled after a cubic fit, as given by the solid line. Therefore, the stiffness has a minimum static value at around the middle of the displacement range, and after this point, the stiffness increases, showing a hardening behaviour for larger deformations. Figure 1c demonstrates this effect. This is useful since an isolator working around this zone will exhibit a low natural frequency, hence good isolation properties. When subjected to cyclic loading, the restoring force on the isolator is hysteretic. The loops in Fig. 1d were obtained by applying a sinusoidal displacement and recording the restoring force. Each loop corresponds to a value of the input displacement. These loops clearly show the nonlinear hardening behaviour. Moreover, these isolators can exhibit different behaviours for different configurations, i.e. shear, roll or tension-compression, and the loops are not always symmetric, depending upon these load conditions. The first step into the analysis of the shock response of such isolators is to study the potential of using the Bouc-Wen model for the response prediction. The following sections evaluate this possibility discussing several shock response parameters for different hardening and softening scenarios.

2.1 Bouc-Wen Model of Hysteresis

The Bouc-Wen model of hysteresis is defined by the following set of differential equations, governing the motion of a single-degree-of-freedom (SDOF) system [12]:

$$\ddot{x} + 2\zeta\omega_n\dot{x} + (1-\gamma)\omega^2 z = u(t) \tag{1}$$

$$\dot{z} = A\dot{x} - \alpha|\dot{x}|z|z|^{n-1} - \beta\dot{x}|z|^n \tag{2}$$

where x, $\dot{x}$ and $\ddot{x}$ represent displacement, velocity and acceleration, respectively, z gives the hsyteretic restoring force, $u(t)$ is the external excitation, γ is the post-yield to pre-yield ratio, and the parameters A, α and β regulate the size of the hysteresis loop and n the smoothness. Through proper selection of the parameters, a wide variety of loops can be represented, as well as energy dissipation characteristics, as depicted in Fig. 2, which presents different loops for hardening and softening behaviour. For instance, when $\alpha = 0.85$ and $\beta = -0.15$, the behaviour is softening, as depicted by the loop in Fig. 2e. On the other hand, when $\alpha = 0.15$ and $\beta = -0.85$, the system presents hardening characteristics as can be seen in Fig. 2i.

2.2 Shock Excitation

Mechanical shock is defined as a short, i.e. transient, excitation, with a duration usually less than twice the natural period of the system. The standard approach for mathematical modelling and predicting the shock response is to consider a pulse function as external excitation, such as a half-sine, rectangular, triangular or another pulse function. The choice of pulse function depends upon the particular situation, and it can represent a variety of scenarios, i.e. a base displacement, base acceleration, force input, etc. For instance, Fig. 3 depicts a SDOF model under a half-sine excitation in the form of an acceleration input. The plot defines several pulses of different durations.

The equation of motion of the system when subjected to a half-sine acceleration pulse of maximum amplitude ξ_p and acceleration response of the system given by υ is

$$\ddot{x} + 2\zeta\omega_n\dot{z} + \omega_n^{\ 2} z = \xi_p \sin\frac{\pi t}{\tau} \tag{3}$$

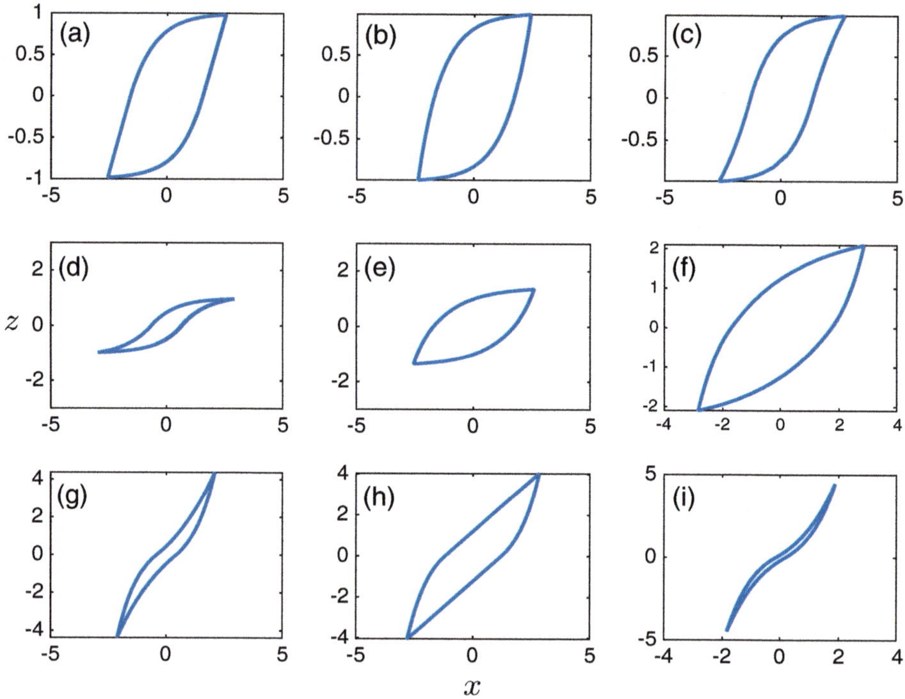

Fig. 2 Hysteresis loops resulting from different parameters of the Bouc-Wen model. (**a**) $\alpha = -0.75$, $\beta = 0.25$; (**b**) $\alpha = 1$, $\beta = 0$; (**c**) $\alpha = 0.25$, $\beta = 0.75$; (**d**) $\alpha = 0.05$, $\beta = 0.95$; (e) $\alpha = 0.85$, $\beta = -0.15$; (**f**) $\alpha = 0.8$, $\beta = -0.2$; (**g**) $\alpha = 0.3$, $\beta = -0.7$; (**h**) $\alpha = 0.5$, $\beta = -0.5$; (**i**) $\alpha = 0.15$, $\beta = -0.85$

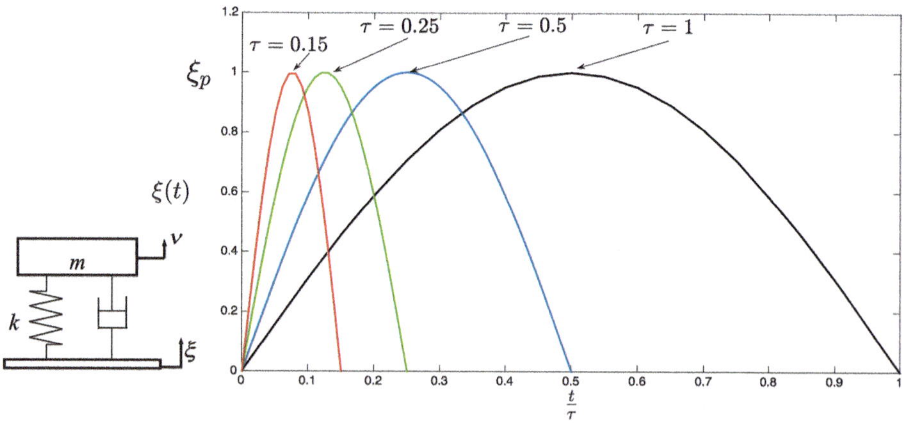

Fig. 3 Representation of a half-sine pulse as shock excitation, considering different values of the duration τ of the pulse

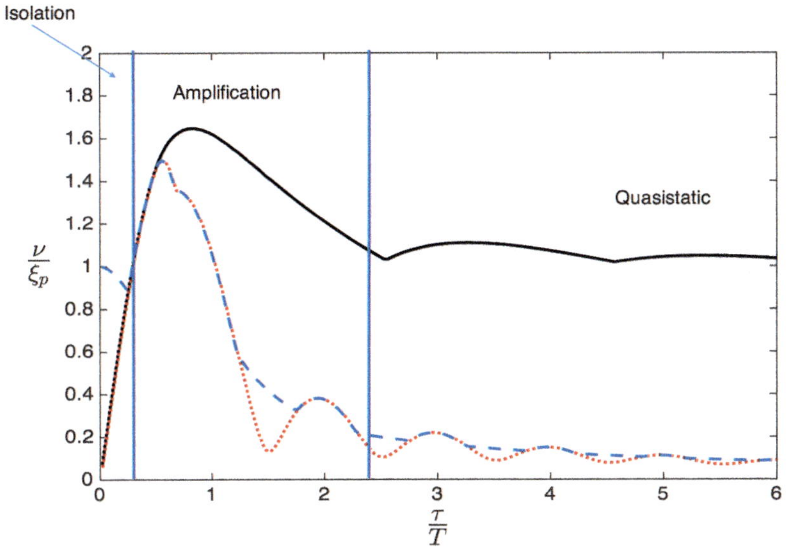

Fig. 4 Shock response spectra of a lightly damped SDOF system

where $z = x - y$ and represents the relative motion between the base displacement y and the mass displacement x. Figure 2 depicts a SDOF model under a versed sine excitation in the form of an acceleration input. The plot defines several pulses of different durations. The response of the system is calculated in two stages, namely, the forced response during the input and the subsequent free or residual vibration. In order to present the shock response in a condensed format, a useful tool is the shock response spectra (SRS) which is a plot representing the normalised shock response, either maximum, residual or relative, as a function of the ratio of the duration of the shock τ and the natural period of the system T. The nature of the response parameter can represent absolute acceleration, displacement or relative displacement depending upon the nature of the input function. There are three typical zones in the SRS depending on the relative duration of the pulse. When the pulse is short compared to the natural period, it is said to be impulsive, and the response is smaller than the input amplitude, resulting in isolation from the impact. Pulses of duration similar to the natural period result in an amplified response. Finally, for longer pulses, the excitation is no longer transient but quasistatic, and it follows closely the input. These three scenarios can be observed in Fig. 4, which gives a SRS corresponding to a lightly damped SDOF system under a half-sine pulse excitation. The continuous line represents the maximum response at any time, sometimes called maximax. The relative and residual responses are given by the broken line and the dotted line, respectively.

3 Shock Response of Hysteretic Systems

For the purpose of this investigation, a half-sine excitation of different durations and amplitudes was considered. The effect of the parameters α and β is considered first. As the number of possible combination of parameters is quite high, the other parameters are considered fixed following the restraints recommended for the model, i.e. $A = 1$. To obtain a complete analysis would be very extensive, and as normally the parameters are identified for a particular case, the choice of parameters presented here gives a good overview of the response for different hysteretic behaviours, i.e. hardening and softening systems. Then, the effect of the input amplitude is studied, which is expected to differ significantly due to the nonlinear properties. For the analysis of the system, the response is divided into two parts, i.e. the forced vibration where the external excitation has the form of a half-sine base acceleration, given by

$$\xi(t) = \frac{-\ddot{u}}{\omega_n^2} = \xi_p \sin\frac{\pi t}{\tau} \tag{4}$$

After the pulse ends, when $t = \tau$, the system is on free residual vibration. The response of the system was calculated by numerical integration in MATLAB. The maximum acceleration due to the pulse is found and presented in the form of a maximax SRS, and then the relative motion and residual responses are calculated and presented as SRS as well.

3.1 *Response to Half-Sine Pulse*

The maximax response for the different parameters considered is presented in Fig. 5. The response is normalised considering the maximum input amplitude ξ_p. Each curve is given for a particular pair of α and β values as —— $\alpha = 0.5$, $\beta = 0.5$; -·-·- $\alpha = 0.7$, $\beta = -0.3$; ··· $\alpha = 0.3$, $\beta = -0.7$; --- $\alpha = 0.85$, $\beta = -0.15$; and + $\alpha = 0.15$, $\beta = -0.85$. The response of the linear system is included for comparison and given by the dot markers. It is easily seen how the response of the system is mainly affected after the isolation or impulsive zone. This means that for impacts of short relative duration, the nonlinearity of the Bouc-Wen model has no effect on the response when compared to the linear approach, regardless of the nature of the nonlinearity. However, the response is greatly changed at the amplification region. The response of the systems with softening characteristics is increased as for the cases when $\alpha = 0.5$, $\beta = 0.5$; $\alpha = 0.85$, $\beta = -0.15$; and $\alpha = 0.7$, $\beta = -0.3$. In contrast, the hardening properties cause a smaller response when $\alpha = 0.15$, $\beta = -0.85$ and $\alpha = 0{:}3$, $\beta = -0.7$.

For the case of the relative response, the SRS depicted in Fig. 6 presents the several cases studied. As before, the hardening Bouc-Wen model provides a slight smaller response during amplification, whilst having a higher response for the softening parameters. For the impulsive zone, the relative motion tends to be similar to

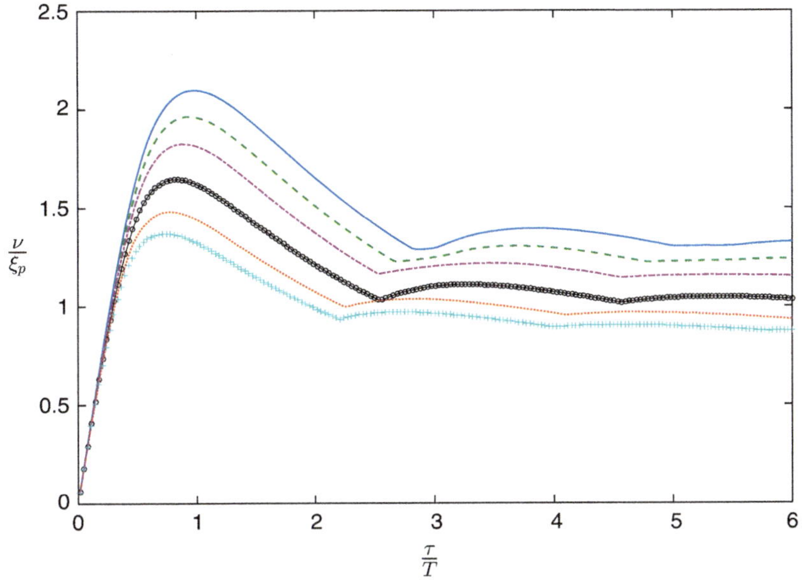

Fig. 5 Shock response spectra for maximum absolute response for a base acceleration input. Vertical axis represents normalised response; horizontal axis gives period ratio (— $\alpha=0.5$, $\beta=0.5$; - · - $\alpha=0.7, \beta=-0.3$; · · · $\alpha=0.3, \beta=-0.7$; - - - $\alpha=0.85, \beta=-0.15$; + $\alpha=0.15, \beta=-0.85$; ○ linear system)

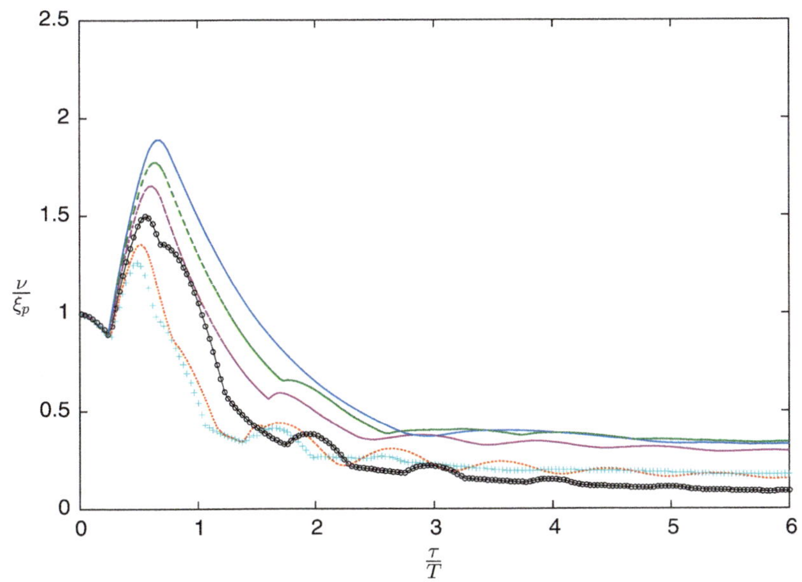

Fig. 6 Shock response spectra for maximum relative motion response for a base acceleration input. Vertical axis represents normalised response; horizontal axis gives period ratio (— $\alpha=0.5$, $\beta=0.5$; - · - $\alpha=0.7, \beta=-0.3$; · · · $\alpha=0.3, \beta=-0.7$; - - - $\alpha=0.85, \beta=-0.15$; + $\alpha=0.15, \beta=-0.85$; ○ linear system)

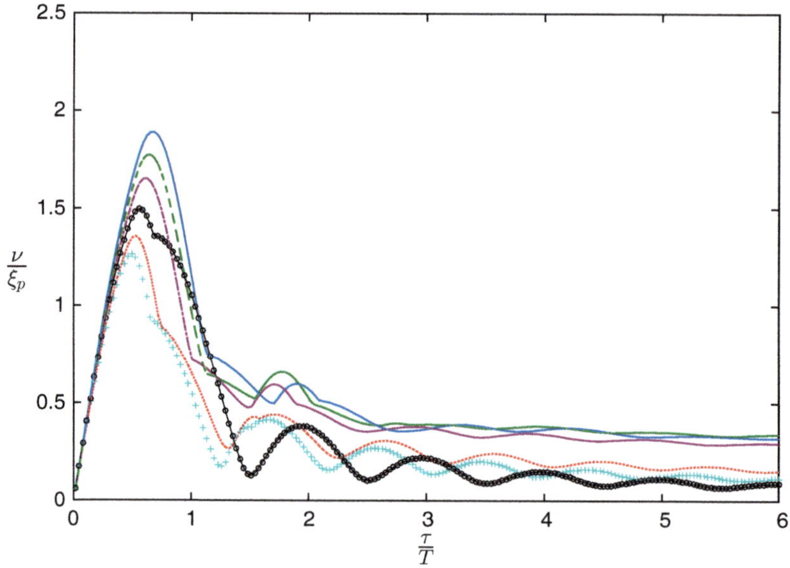

Fig. 7 Shock response spectra for maximum absolute residual for a base acceleration input. Vertical axis represents normalised response; horizontal axis gives period ratio (—— α=0.5, β=0.5; - · - α=0.7, β= −0.3; · · · α=0.3, β= −0.7; - - - α=0.85, β= −0.15; + α=0.15, β= −0.85; O linear system)

the shock amplitude, due to the absolute response being small for very short inputs, which is the same effect observed in the linear system.

Finally, the residual response is presented in Fig. 7. In this case, a similar trend is observed, where the response is the same for very short pulses when different parameters of the model are considered, increasing in the amplification region when the system is softening and decreasing when the system is hardening.

Although the SRS presented for the different parameters studied offer condensed information of the response of the system, sometimes it is useful to look at the time response for particular cases. Fig. 8 presents four situations for period ratios of $\frac{\tau}{T}$, that is, 0.25, 0.5, 1 and 2. For these plots, the Bouc-Wen parameters selected represent a hardening system, i.e. - - - α=0.15, β=−0.85, and a softening system, for · · · α=0.05, β=0.95, whilst the continuous line represents the response of the linear system, and the bold line depicts the half-sine pulse. The effects explained before are readily observed in these plots. When the pulse is of short relative duration, i.e. $\frac{\tau}{T} = 0.25$ and $\frac{\tau}{T} = 0.5$, for Fig. 8a and c, the response occurs after the pulse has ended, i.e. the excitation is impulsive. A slight increase in performance isolation is observed when the system is hardening, whereas the isolation worsens when softening. The most notorious effect occurs when $\frac{\tau}{T} = 1$, in Fig. 8b, for which the softening system presents a considerable response increase and the opposite effect for the hardening model.

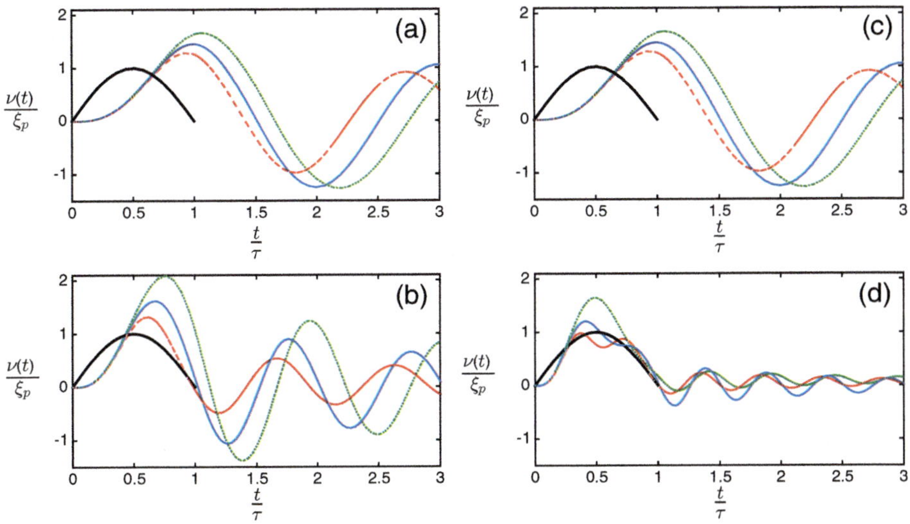

Fig. 8 Time response of absolute acceleration of the Bouc-Wen model, considering different values of the parameters α and β. Half-sine shock pulse depicted by bold line (a) $\frac{\tau}{T}=0.25$, (b) $\frac{\tau}{T}=0.5$, (c) $\frac{\tau}{T}=1$, (d) $\frac{\tau}{T}=2$. (- - - α=0.15, $\beta=-0.85$; · · · α=0.05, $\beta=-0.95$; — linear system)

On the other hand, for large shocks, whilst the hardening effect produces a very small difference compared with the linear model, the softening response is still considerably larger, as presented in Fig. 8d when $\frac{\tau}{T}=2$.

3.2 *Effect of Input Amplitude*

One of the main concerns of the nonlinearity of the systems studied in this work is the effect of the input amplitude, as the common assumptions of a linear system are not valid, i.e. superposition and input-output ratio. In order to assess the effect of the input amplitude, in this section, numerical results are presented considering increments of the input amplitude from 1 to 10 g in 1 g steps. Figure 9a–c was calculated for hardening parameters $\alpha=0.15$ and $\beta=-0.85$, and Fig. 9d–f for softening parameters $\alpha=0.85$ and $\beta=-0.15$. Each case is presented for maximax, relative and residual response parameters. The response is normalised considering maximum input amplitude as before. The direction of the arrows indicates the effect of increasing the input amplitudes. When comparing the hardening with the softening cases, a further decrease in the normalised response for the former and an increment for the latter can be easily observed. The results of the nonlinearity effects are easily observed since the constant input-output ratio of linear system no longer holds on the observed results. The hardening system shows improved isolation properties as the input amplitude increases, and this effect is similar for the maximax and residual

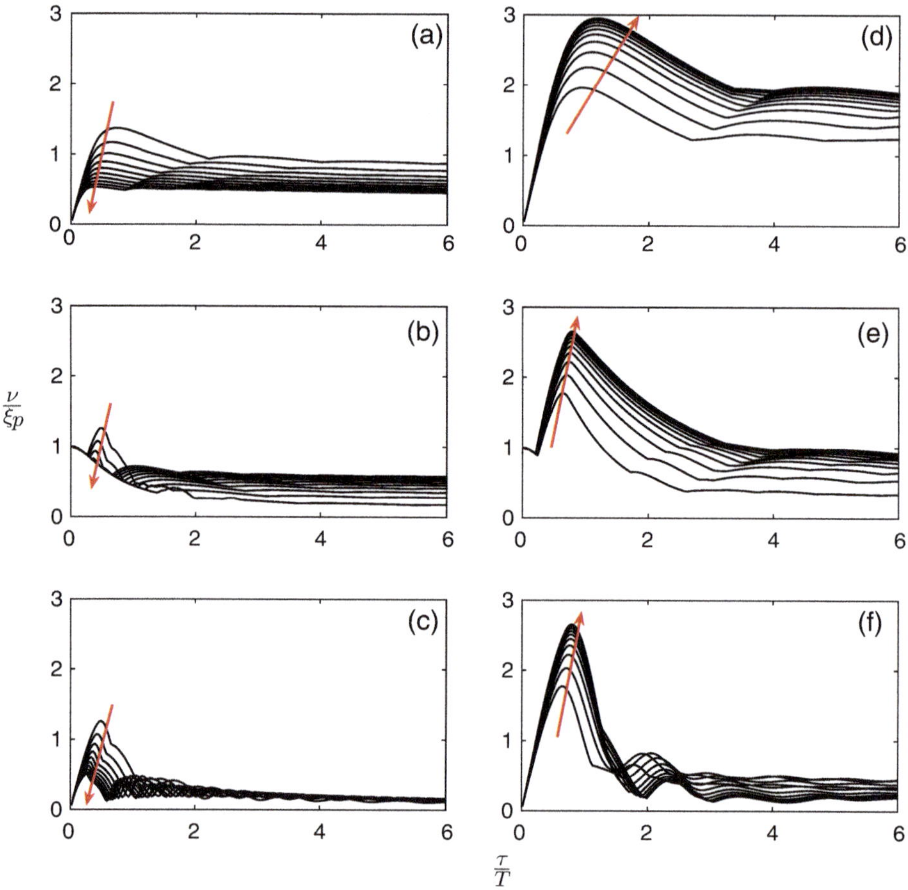

Fig. 9 Effect of input amplitude on the shock response spectra of the Bouc-Wen model for selected hardening and softening parameters. (**a**) Hardening, maximax; (**b**) hardening, relative; (**c**) hardening, residual; (**d**) softening, maximax; (**e**) softening, relative; (**f**) softening, residual. The direction of the arrows shows the effect of increasing the input amplitude

response. This improvement in isolation is more pronounced around the amplification region of the SRS, i.e. when the pulse duration is similar to the natural period of the system. However, the maximum relative response tends to stabilise at one regardless of the input amplitude. This is explained considering that isolation is produced by deflection in the elastic element. The maximum relative response is unitary for very short pulses, where isolation is obtained. Thus, relative response is very similar to the pulse amplitude as the maximum response occurs after the pulse has finished. On the other hand, the softening system shows a nonlinear increase in the normalised response as the input increases thus affecting the isolation performance. The effect is more prominent around the amplification region.

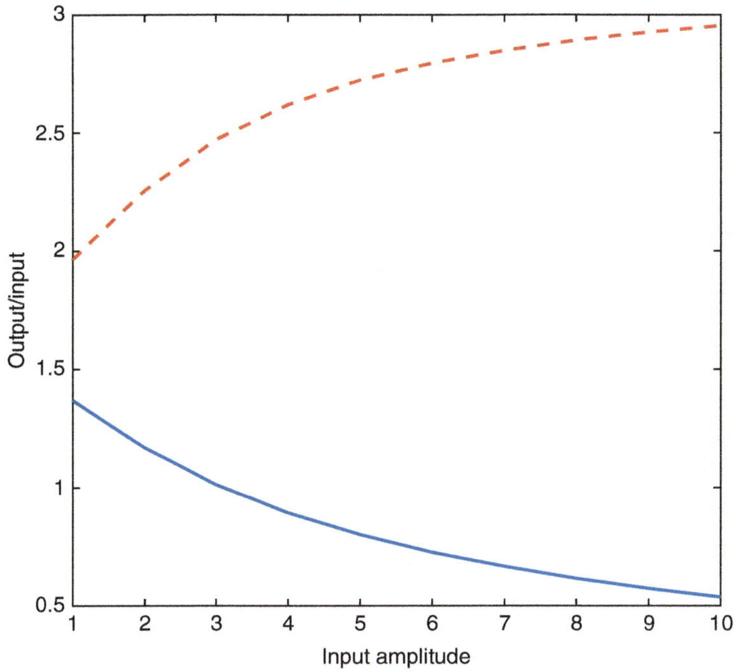

Fig. 10 Normalised maximax shock response as a function of the input amplitude for the Bouc-Wen model for selected hardening and softening parameters (- - - softening system, —— hardening system)

Moreover, relative response is considerably increased around the amplification zone, resulting from the much higher response compared to the input. A similar effect is observed in the residual response.

The previously explained behaviour can be easily appreciated in Fig. 10. This plot presents the effect of the maximum normalised response around the amplification region as a function of the input amplitude. The dashed line represents the softening system, whilst the continuous line depicts the hardening system. Only maximax response is presented since the effect in the other parameters is similar. It can be observed how the response for the hardening system decreases gradually and asymptotically as the input increases, whilst the opposite effect occurs in the softening system.

3.3 Discussion

Previous studies have demonstrated theoretically and experimentally the feasibility and advantages of using a nonlinear hardening isolation system for shock isolation [19, 20]. In general, the performance in absolute response parameters is increased, i.e. maximum displacement and acceleration, but the relative motion is affected. This is expected, as the mechanism of shock isolation relies on the deformation exerted in the elastic element. The model considered here demonstrates a similar behaviour, in which better isolation can be obtained through the use of a nonlinear hardening system. In fact, the hardening system that provides better isolation properties has a hysteresis loop with low damping, i.e. a small area and hardening stiffness (refer to Fig. 2i) similar to the cubic stiffness approximations explored previously by [19]. However, this study considers a different approach which might be more suitable for systems with hysteretic damping, such as viscoelastic mounts, or dry friction isolators when subjected to shock inputs. The Bouc-Wen model might be a more complicated approach due to the number of parameters involved and the semiempirical nature, ruling out the possibility of closed form solutions, but it is justified for highly damped systems where other models only consider an equivalent viscous damping or loss factor. Although the effect of damping is not considered in this analysis, it can be included into the Bouc-Wen model through proper selection of the values defining the loop. A particular type of isolators that can be properly modelled considering this approach are friction-based isolators or cable mounts. However, one of the most important aspects to consider for further study is the identification of Bouc-Wen parameters for particular isolation systems. There are several approaches for parameter identification, such as genetic algorithms, particle swarm, etc. Moreover, these isolators present different hysteretic behaviours depending upon the direction of loading, i.e. can show symmetric or asymmetric loops, which for the latter case a modified Bouc-Wen model would be more suitable. However, the theoretical results presented in this paper give an overview of the shock response to be expected in this type of isolators, which represent a challenging and interesting task for future research.

4 Conclusions

A theoretical analysis of the shock response of a SDOF isolator with hysteresis was presented, based on the Bouc-Wen model of hysteresis. The Bouc-Wen model is extensively used to represent system with hysteretic behaviour, such as structural damping, or systems with nonlinear properties, and depending upon its parameters, a great variety of loops can be represented. In this chapter, the isolator was subjected to shock pulses of different durations and amplitudes, considering hysteretic systems with softening and hardening characteristics. It was found that a system with hardening behaviour has a better shock isolation compared with a softening system. The effects of input amplitude demonstrate the nonlinear response; as the shock input increases, the input-output ratio decreases, providing better isolation for larger displacements. This approach could be useful for the modelling and prediction of shock response in nonlinear isolators commonly used in shock isolation, such as wire rope springs, which is suggested for further work.

References

1. Mercer, C. A., & Rees, P. L. (1971). An optimum shock isolator. *Journal of Sound and Vibration, 18*(4), 511–520.
2. Snowdon, J. C. (1961). Response of nonlinear shock mountings to transient foundation displacements. *The Journal of the Acoustical Society of America, 33*(10), 1295–1304.
3. Molyneux, W. G., & Aeronautical Research Council. (1956). *Supports for vibration isolation* (p. 8). London: H.M. Stationery Office.
4. Kutscha, D., & Eshleman, R. L. (1969). *shock isolation elements testing for high input loadings, Mechanical shock isolation elements* (Vol. 3, p. 353). Chicago IL: IIT Research Inst.
5. Eshleman, R. L. (1972). Dynamic response of a ring spring. In *The shock and vibration bulletin* (p. 7). Washington, DC: The Shock and Vibration Information Center.
6. Tinker, M. L., & Cutchins, M. A. (1992). Damping phenomena in a wire rope vibration isolation system. *Journal of Sound and Vibration, 157*(1), 7–18.
7. Cutchins, M. A., Cochran, J. E. J., Kumar, K., Fitz-Coy, N. G., & Tinker, M. L. (1988). Analysis of coils of wire rope arranged for passive damping. *Recent Advances in Structural Dynamics, 1*, 469.
8. Gerges, R. R., & Vickery, B. J. (2003). Parametric experimental study of wire rope spring tuned mass dampers. *Journal of Wind Engineering and Industrial Aerodynamics, 91*, 1363–1385.
9. Hogea, D., & Bausic, F. (2005). *Vertical static tests of the wire rope isolators*. Romania: Technical University of Civil Engineering of Bucharest.
10. Ismail, M., Ikhouane, F., & Rodellar, J. (2009). The hysteresis Bouc-Wen model, a survey. *Archives of Computational Methods in Engineering, 16*(2), 161–188.
11. Charalampakis, A., & Koumousis, V. (2008). Identification of Bouc-Wen hysteretic system by a hybrid evolutionary algorithm. *Journal of Sound and Vibration, 314*(3), 571–585.
12. Li, H.-g., & Meng, G. (2007). Nonlinear dynamics of a SDOF oscillator with Bouc–wen hysteresis. *Chaos, Solitons & Fractals, 34*(2), 337–343.
13. Ni, Y. Q., Ko, J. M., & Wong, C. W. (1998). Identification of non-linear hysteretic isolators from periodic vibration tests. *Journal of Sound and Vibration, 217*(4), 737–756.
14. Gerges, R. R. (2008). Model for the force–displacement relationship of wire rope springs. *Journal of Aerospace Engineering, 21*(1), 1–9.
15. Di Massa, G., Pagano, S., Rocca, E., & Strano, S. (2013). Sensitive equipments on WRS-BTU isolators. *Meccanica, 48*(7), 1777–1790.
16. Wang, H.-X., Gong, X.-S., Pan, F., & Dang, X.-J. (2015). Experimental investigations on the dynamic behaviour of O-type wire-cable vibration isolators. *Shock and Vibration, 2015*, 1–12.
17. Carboni, B., Lacarbonara, W., & Auricchio, F. (2015). Hysteresis of multiconfiguration assemblies of nitinol and steel strands: Experiments and phenomenological identification. *Journal of Engineering Mechanics, 141*(3), 1–16.
18. Guzmán-Nieto, M., Tapia-González, P. E., & Ledezma-Ramírez, D. F. (2015). Low frequency experimental analysis of dry friction damping in cable isolators. *Journal of Low Frequency Noise, Vibration and Active Control, 34*(4), 513–524.
19. Tang, B., & Brennan, M. J. (2014). On the shock performance of a nonlinear vibration isolator with high-static-low-dynamic-stiffness. *International Journal of Mechanical Sciences, 81*, 207–214.
20. Ledezma-Ramirez, D. F., Ferguson, N. S., Brennan, M. J., & Tang, B. (2015). An experimental nonlinear low dynamic stiffness device for shock isolation. *Journal of Sound and Vibration, 347*, 1–13.

Part II
Corrosion in Aerospace Materials

Effects of Heat Treatments in Corrosion Behavior of HVOF-Sprayed Ni-Based Alloy Coatings

J.A. Cabral-Miramontes, R. Rocha Reséndez, C. Gaona Tiburcio, P. Zambrano-Robledo, C. Poblano Salas, and F. Almeraya Calderon

Abstract It is well known that sprayed Ni-based alloy coatings show good corrosion resistance. They have good wear resistance after adding W and Mo elements to the alloy. High-velocity oxygen-fuel (HVOF) spraying is a thermal spray technique well matched to obtain corrosion- and wear-resistant coatings. Post-spray treatments such as laser remelting, shot peening, HIP, sealing, and furnace heat treatments are also used to improve the density and homogeneity of sprayed coatings. The corrosion properties of Ni-based alloy coatings processed by HVOF thermal spraying were studied as a function of annealing heat treatment at 850 and 950 °C in atmosphere of vacuum and air. Immersion tests in HCl, NaOH, and water media with various electrochemical techniques such as linear polarization resistance (LPR) and potentiodynamic anodic polarization curves (PAPC) were employed to perform such evaluation. The microstructural characterization of the as-sprayed coatings was analyzed by scanning electron microscopy (SEM). The results show that when the heat treatment temperature increases, the corrosion current density gradually decreases. This implies that thermally treated coatings increase its corrosion resistance relative to the coating of the substrate and heat treatment. Finally, it was found that a heat treatment at 950 °C in vacuum atmosphere improves the corrosion resistance of Ni-based coating applied by HVOF thermal spray.

Keywords Corrosion • HVOF • Heat treatment • Thermal spray

J.A. Cabral-Miramontes (✉) • R. Rocha Reséndez • C. Gaona Tiburcio
P. Zambrano-Robledo • F. Almeraya Calderon
Universidad Autónoma de Nuevo León, UANL. Facultad de Ingeniería Mecánica y Eléctrica, FIME. Centro de Investigación e Innovación en Ingeniería Aeronáutica, CIIIA, Aeropuerto Internacional del Norte, Carretera a Salinas Victoria Km. 2.3, Apocada 66600, NL, Mexico
e-mail: josecabral.miramontes80@gmail.com

C. Poblano Salas
Centro de Tecnología Avanzada (CIATEQ A. C.), Av. Manantiales 23-A, Parque Industrial Bernardo Quintana, El Marques, Querétaro, Mexico

P. Zambrano-Robledo et al. (eds.), *Proceedings of the Symposium of Aeronautical and Aerospace Processes, Materials and Industrial Applications*, DOI 10.1007/978-3-319-65611-3_5

1 Introduction

The thermal spray high-velocity oxygen-fuel (HVOF) is a commonly used technique for bonding coatings to a surface, improving its characteristics [1–3]. The powder particles are injected to a high-velocity hot gas generated by combustion. Oxygen and fuel gas flow at high pressure producing very high particle velocities at low temperatures compared to other thermal processes, such as spray or vacuum plasma air [4].

The HVOF is able to produce coatings resistant to high temperatures and wear and lower losses of carbides, resulting in a material with high hardness [5]. Wear resistance and toughness are characteristics that have the ability to tolerate aqueous corrosion; also, it creates a low-porosity coating, adhering better to the substrate [6–9]. Temperatures around 2000 °C and speeds between 600 and 800 m/s, depending on the powder used [10]. These conditions are appropriate, since it is an ideal for corrosion resistance coating [11].

Nickel-based coatings whose main feature that support high-temperature corrosion and wear, refractory materials such as W and Mo and metalloids such as B and C as base powders was used [12, 13]. There are several reports related to nickel-based spray and other metal alloys resistant to corrosion, where coatings are obtained with low-porosity and crystalline microstructures [7–9, 14]. There are investigations where it was found that the coatings Ni-Cr-W-Mo have strength similar to that of the substrate without generating corrosion products, also finding evidence of exfoliation and laminar peeling when the coating system was heat-treated in high vacuum [15–17]. Heat treatment furnaces are used to improve the density and homogeneity of the coatings applied by thermal spraying [18, 19].

The objective of this study is to evaluate the corrosion resistance of the coating Ni-based alloy deposited by HVOF thermal spray using a substrate of the 304 stainless steel and applying heat treatment in air compared to high vacuum conditions.

2 Experimental Procedure

2.1 Materials

To obtain coatings by HVOF, the equipment hybridizes DJH 2700 which was employed as a base powder for spraying an alloy Ni-Cr-W-Mo-Cu-B-C-Fe, whose characteristics are included in Table 1. Bulk density powder was measured following

Table 1 Characteristics of the powder Ni-based alloy

Chemical composition[a]	Bal Ni, 20% Cr, 10% W, 9% Mo, 4% Cu, 1% C, 1% B, 1% Fe	*Apparent density*	3.382 g/cm^3
Particle size	−53 + 11 μm (mesh −270 + 11 μm)	*Flow rate*	4.24 s/50 g

[a]Weight percent

the guidelines of ASTM B212 [20], while the flow was measured using the ASTM B964 [21]. All coatings were sprayed on samples AISI/SAE 304 stainless steel substrate with folowing dimensions of 5 × 5 × 0.5 cm. Cleaning conditions of the surfaces of the samples were prepared by sand blasting (Al_2O_3 Grade G-20), at a distance of 0.2 m and an air pressure of 4 bar, to achieve a surface roughness (Ra) of 5–10 μm to enhance the mechanical adhesion between the substrate and the projected particles.

2.2 *HVOF Spraying*

Deposition conditions of the powder Ni-based alloy with respect to pressures and flows are suggested in the literature [12], so that it could be a reference for the manufacture of coating. The conditions used for the coating are included in Table 2. For microstructural characterization of the coatings, cross section coated samples were performed. The microstructure of the powder and the coatings were studied using scanning electron microscopy JEOL JSM 6510 V to quantify contaminant particles in the interface, determining the amount of oxides, porosity, and unmelted particles present in the coating.

2.3 *Heat Treatments*

Two thermal treatments in controlled atmosphere furnace with high vacuum and air at temperatures of 850 and 950 °C were conducted with cooling in the same furnace and maintaining the atmospheres. In Fig. 1, we can see a graph representing the heat treatment coating to 850 and 950 °C.

2.4 *Electrochemical Tests*

Electrochemical tests were carried out with a potentiostat-galvanostat/ZRA Gill AC-1. Technique linear polarization resistance (LPR) ASTM G59 standard [22] was used, with parameters ±20 mV polarization, with respect to the corrosion potential (Ecorr), with a scanning rate of 10 mV/min. Potentiodynamic polarization curves

Table 2 HVOF spraying parameters

Spraying distance	229 mm	*Propane flow*	25 l/min
Manipulator speed	1.5 m/s	*Airflow*	30 l/min
Oxygen pressure	150 psi	*Gas haul*	Nitrogen
Propane pressure	90 psi	*Pressure gas rush*	150 psi
Air pressure	100 psi	*Gas flow carrying*	28.5 l/min
Oxygen flow	37 l/min	*Powder feed*	45 g/min

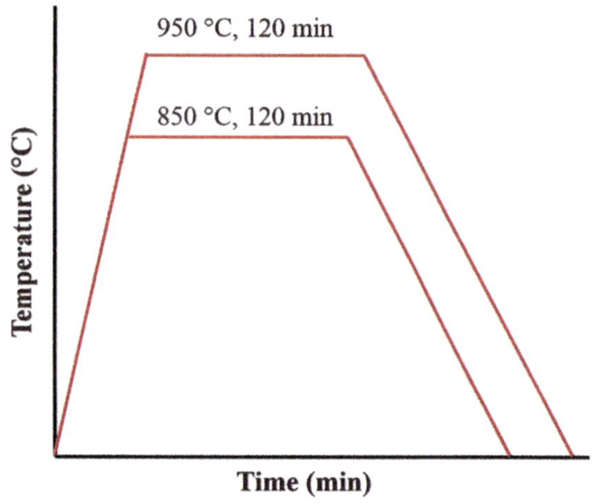

Fig. 1 Heat treatments to Ni-based alloy at 850 and 950 °C in atmosphere of vacuum and air

Fig. 2 Electrochemical cell used in this work. (*1*) Working electrode, (*2*) SCE reference electrode, and (*3*) platinum auxiliary electrode

(PPC) were initiated with potential −300 mv relative to the corrosion potential, a sweep rate of 60 mV/min, and final potential of 1500 mV according to standard ASTM G5 [23]. These experiments were performed in a pitting cell using an array of three electrodes: the reference electrode [saturated calomel electrode (SCE)], a platinum auxiliary electrode, and working electrode coating Ni-based alloy (304 stainless steel used as substrate). This electrochemical arrangement is shown in Fig. 2. The coatings were in contact with solutions of 0.5 N hydrochloric acid (HCl), 0.5 N sodium hydroxide (NaOH), and distilled water at room temperature.

3 Results

3.1 Coating Microstructure

Figure 3 shows the microstructures obtained by scanning electron microscopy (SEM) representative cross section of the coating type Ni-based alloy that are presented once deposited by the technique of high-velocity oxygen-fuel. It can be observed that the coating is homogeneous, showing little porosity, lamellar structure typical of coatings applied by thermal spray, both low oxide contaminations as coating-substrate interface.

3.2 Linear Polarization Resistance

Table 3 shows the results of linear polarization resistance of the Ni-based alloy coatings with various thermals and subjected to different treatments that are presented in aqueous media. It can be seen that the behavior of heat treatment in air is much more negative than the thermal treatments performed in a vacuum. Its more active, the coating without heat treatment and treated in vacuum atmosphere presented

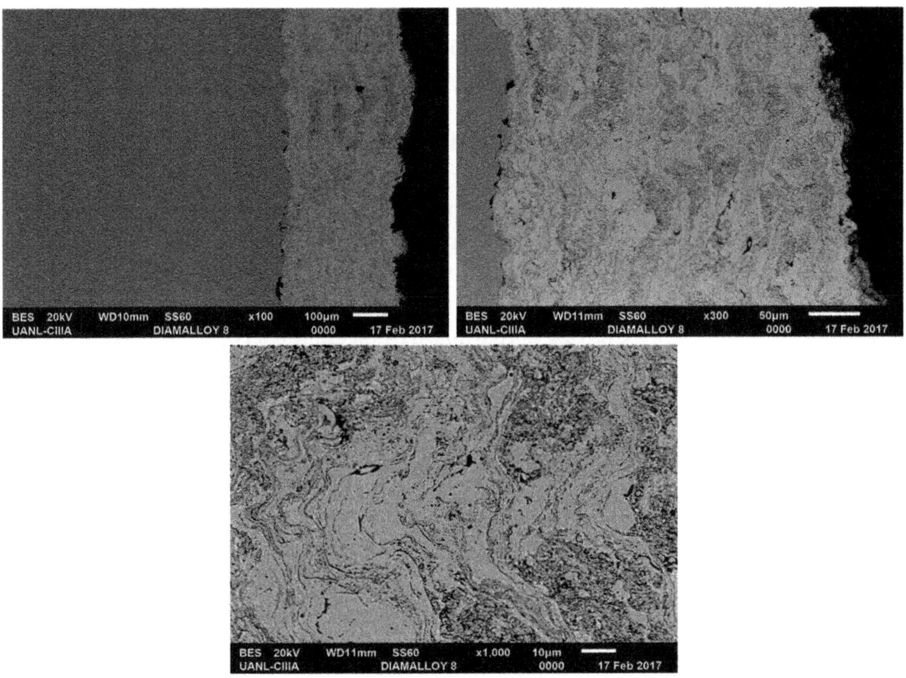

Fig. 3 Microstructural transversal section of Ni-based alloy coating by HVOF thermal spray

Table 3 Corrosion potential (Ecorr), lineal polarization resistance (RPL), and corrosion current density (jcorr) of Ni-based alloy coatings with heat treatments

Solution	Sample	Ecorr (mV vs. ECS)	LPR (Ω-cm^2)	jcorr (mA/cm^2)
H_2O	NiCoWMo	3.09	438,450	5.74E-4
	V850	−72.50	140,210	20.8 E-4
	V950	−157.84	157,070	18.6 E-4
	A850	−87.95	882,530	2.85 E-4
	A950	−415.15	149,180	1.89 E-4
HCl	NiCoWMo	−102.48	395.32	6.37E-2
	V850	−218.55	485.21	6.01 E + 0
	V950	−225.48	12,040.00	2.42 E-2
	A850	−79.72	194.6	1.45 E + 0
	A950	12.65	196.78	1.43 E + 0
NaOH	NiCoWMo	−272.19	43,213	5.82E-4
	V850	−152.23	9097.5	3.10 E-4
	V950	−401.25	52,862	0.34 E-4
	A850	−173.29	11,930	2.44 E-4
	A950	−129.71	18,151	1.60 E-4

corrosion current density values very similar corrosion rate at least an order of magnitude in mass HCl and NaOH (7.33E + 00 6.01E + 00 and respectively). Note that when the coating tested in the hydrochloric acid solution is in constant activation, this being indicative of the presence of the chloride ion, it could modify the coating properties. In papers presented by Zhao et al. [24], it is argued that unmelted particles during the thermal spraying step, pores, inclusions, and lamellar structure microcracks cause further degradation of thermal spray coatings applied. However, in this study, the coating has practically evaluated these characteristics being the main reason why the coating material behaves at speeds similar corrosion to base material. It can also be seen that the most aggressive means coating is the HCl, and heat treatment at 950 °C in vacuum atmosphere is the one that provides the best results in the three means of evaluation. The behavior of thermal-treated coatings is directly related to the change of morphology and porosity-decreased coating due to heat treatments, which has a beneficial effect on corrosion resistance. The results obtained in this study agree with those reported by [16] claiming that similar chemical composition coatings applied by HVOF interconnected porosity are very low, and therefore corrosion penetration of the coating is not the main concern.

3.3 Potentiodynamic Anodic Polarization Curves

Figure 4 shows the behavior of the coating in distilled water under two different atmospheres and the two heat treatments and without heat treatment coating. In the graphic, we can observe that vacuum has greater corrosion resistance, at

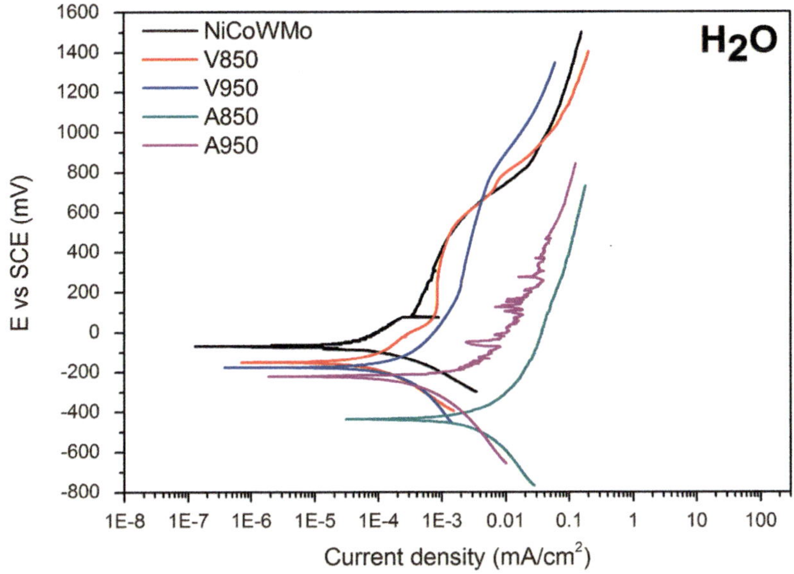

Fig. 4 Potentiodynamic anodic polarization curves for Ni-based alloy coatings in H_2O

850–950 °C. The behavior of the coating at 950 °C in vacuum is similar to the coating without heat treatment. We also note that air has a lower corrosion resistance in both vacuum heat treatment temperatures. It is observed that the coating without heat treatment has a lower current demand, compared to those tests in air, which states that it better resists corrosion without any of the treatments in air. Treatment that resists more corrosion is vacuumed at 950 °C.

In Fig. 5, we can observe the behavior of coating Ni-based alloy in HCl under two different heat treatments in air and vacuum. We can see in Fig. 5 that the heat treatment in air at 850 and 950 °C has lower corrosion resistance than the coating without heat treatment. It is noted that the vacuum coating at 950 °C is the most resistant to corrosion and has a tendency to passivation in the anodic zone.

Finally, in Fig. 6, the two heat treatments in different atmospheres and coating without heat treatment in NaOH solution are compared. With a similar behavior in all conditions, reaching its respective passivation but with certain differences in the passivation current density, treatment in vacuum at 850 °C has a similar corrosion rate without coating heat treatment. For the samples treated in air, the corrosion rate is higher as was observed in the two previous solutions.

It is observed that the cathodic region of all curves exhibits a similar behavior. When the temperature of the heat treatment increases, the corrosion current density gradually decreases. This implies that the heat-treated coatings enhance corrosion resistance in connection with heat treatment of the coating and the substrate. This phenomenon is directly related with reduced porosity and better bonding between the particles, leading to blockage of the penetration route to the corrosive solution, and may be attributed to the homogenization of multiple crystal phases that can be

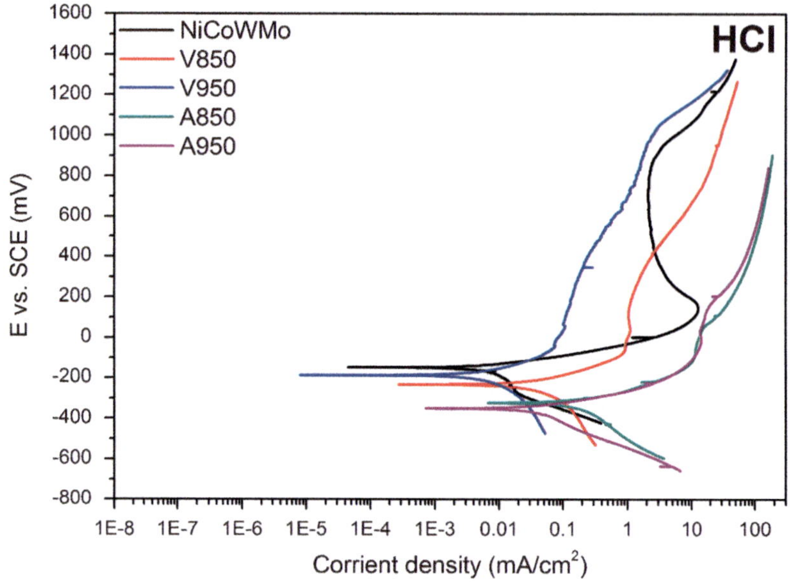

Fig. 5 Potentiodynamic anodic polarization curves for Ni-based alloy coating in HCl

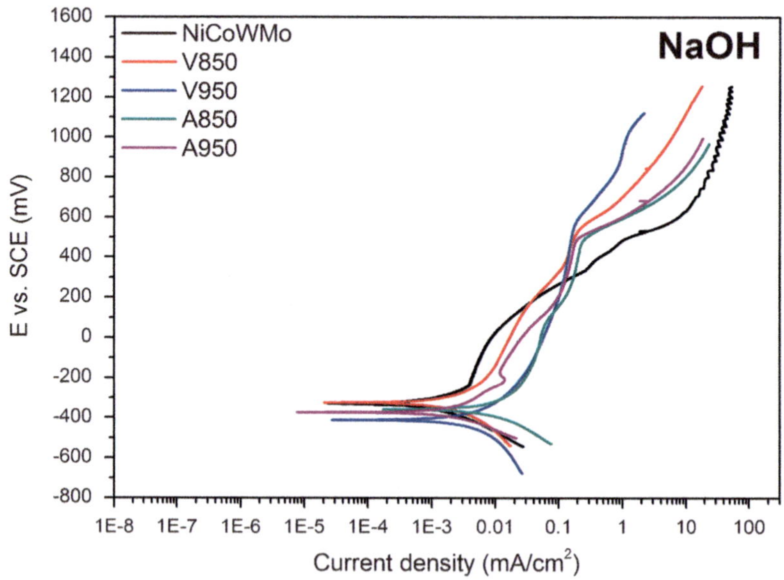

Fig. 6 Potentiodynamic anodic polarization curves for Ni-based alloy coating in NaOH

induced phenomenon phases of galvanic corrosion between the materials without the heat treatment. In the anodic region, the coatings processed at 950 °C exhibit lower passivation current density, approximately an order of magnitude less than coating without heat treatment and a little lower than the coatings processed at 850 °C. In some cases, this type of behavior is attributed to the formation of large amount of precipitates, $M_{23}C_6$ and M_7C_3 type on the surfaces, which are very stable and can be the cause of lower passivation current density in the coating treated with these temperatures [17].

4 Conclusions

- The coatings were obtained with good properties such as low porosity, low oxide content, and little interface contamination, and no presence of unmelted particles was observed.
- The Ni-based alloy coating without heat treatment has a lower corrosion rate than the two heat treatments in air and vacuum.
- Heat treatments in air have a lower corrosion resistance than those treated in vacuum.
- Potentiodynamic polarization curves, which were performed in HCl and NaOH, demonstrate passivation coatings and resists mainly in vacuum at 950 °C.

References

1. Smith, R. W., & Knight, R. (1995). Thermal spraying I: Powder consolidation—From coating to forming. *Journal of Metals, 47*(8), 32.
2. Pawlowsky, L. (2008). *The science and engineering of thermal spray coatings*, second edition, (p. 67). John Wiley and Sons, LTD. ISBN: 978-0-471-49049-4.
3. Marulanda Arevalo, J. L. (2007). *Rociado Termico* (p. 34). Colombia: Universidad Tecnologica de Pereira , UTP.
4. Wirojanupatump, S., Shipway, P. H., & McCartney, D. G. (2001). *Wear, 249*, 829–837.
5. Wang, B. Q., & Shui, Z. R. (2002). The Hot Erosion Behavior of HVOF Chromium Carbide-Metal Cermet Coatings Sprayed with Different Powders. *Wear, 253*, 550–557.
6. Stokes, J. (2008). *The theory and application of the HVOF thermal spray process* (p. 53). Dublin City University. ISBN: 1-87232-753-2, ISSN: 1649-8232.
7. Yuping, W., Pinghua, L., Chenglin, C., Zehua, W., Cao., M., & Junhua, H. (2007). Cavitation erosion characteristics of a Fe-Cr-Si-B-Mn coating fabricated by high velocity oxy-fuel (HVOF) thermal spray. *Materials Letters, 61*(2007), 1867–1872.
8. Goyal, R., Singh, K., & Singla, S. (2014). Wear behaviour of thermal spray coating process on carbon steel. *International Journal of Research in Mechanical Engineering and Technology, 4*(2), 182–185. ISSN: 2249-5762 (on line).
9. Edris, H., McCartney, D. G., & Sturgeon, A. J. (1997). Microstructural characterization of high velocity oxy-fuel sprayed coatings of inconel 625. *Journal of Materials Science, 32*, 863.
10. Davis, J. R. (2004). *Handbook of thermal spray technology* (pp. 54, 52 and 263). United States of America: Prepared under the direction of the Thermal Spray Society Training Committee. ISBN: 0–87170–795-0.
11. Dent, A. H., Horlock, A. J., McCartney, D. G., & Harris, S. J. (2000). Microstructure formation in high velocity oxy-fuel thermally sprayed Ni-Cr-Mo-B alloys. *Materials Science and Engineering, A283*, 242–250.

12. Lee, C. H., & Min, K. O. (2000). Effects of heat treatment on the microstructure and properties of HVOF-sprayed Ni–Cr–W–Mo–B alloy coatings. *Surface and Coatings Technology, 132*, 49–57.
13. Kawakita, J., Fukushima, T., Kuroda, S., & Kodama, T. (2002). Corrosion behaviour of HVOF sprayed SUS316L stainless steel in seawater. *Corrosion Science, 44*, 2561–2581.
14. Knight, R., & Smith, R. W. (1992). In C. C. Berndt (Ed.), *Proceedings of 13th international thermal spray conference, Orlando, FL, USA, 28 May 5 June, 1992* (p. 159). Metals Park, OH: ASM International.
15. Cabral-Miramontes, J. A., Gaona-Tiburcio, C., Almeraya-Calderón, F., Estupiñan-Lopez, F. H., Pedraza-Basulto, G. K., & Poblano-Salas, C. A. (August 2014) Parameter studies on high-velocity oxy-fuel spraying of CoNiCrAlY coatings used in the aeronautical industry". XXIX Mexican Electrochemical Society Congress. Quintana Roo, México.
16. Cabral Miramontes, J. A., Achem Calahorra, J. Y., Pedraza Basulto, G. K., Zambrano Robledo, P., Poblano Salas, C., & Almeraya Calderón, F. (August 2014). Effects of the heat treatments on the corrosion behavior of Ni-Cr-W-Mo alloy coatings applied by HVOF technique. XXIII International Material Research Congress joint with the XIII Congreso NACE México. Quintana Roo, México.
17. Achem Calahorra, J. Y., Gaona Tiburcio, C., Estupiñan López, F. H., Zambrano Robledo, P., Poblano-Salas, C., Almeraya Calderón, F., & Cabral Miramontes, J. A. (October 2014). Resistencia a la Corrosión de Recubrimientos Ni-Co-W-Mo Aplicados por Rociado Térmico HVOF. IX Latin-American Congress of Corrosion (LATINCORR). Medellin, Colombia.
18. Ishida, M., Sasaki, M., Hattori, T., & Nakagishi, Y. (1995). High corrosion resistance of a Ni-Alloy sprayed and Pd-P Plated coating. *Proceedings of ITSC*, 1011.
19. Kishitake, K., Era, H., & Otsubo, F. (1996). Thermal-sprayed Fe-10CM3P-7C amorphous coatings possessing excellent corrosion resistance. *Journal of Thermal Spray Technology, 5*(4), 476.
20. ASTM B212. (2013). *Standard test method for apparent density of free-flowing metal powders using the hall flowmeter funnel*. ASTM annual book of standards, 2013.
21. ASTM B964. (2013). *Standard test methods for flow rate of metal powders using the carney funnel*. ASTM annual book of standards, 2013.
22. ASTM G59-94. (1999). *Standard reference test method for making potentiostatic and potentiodynamic anodic polarization measurements*. ASTM annual book of standards, 1999.
23. ASTM G5-94. (1999). *Standard reference test method for making potentiostatic and potentiodynamic anodic polarization measurements*. ASTM annual book of standards, 1999.
24. Zhao, W., Wang, Y., Dong, L., Wu, K., & Xue, J. (2005). Corrosion mechanism of NiCrBSi coatings deposited by HVOF. *Surface and Coatings Technology, 190*, 293–298.

Electrochemical Study of 17-4 and 17-7 PH Stainless Steels Used in the Aeronautical Industry

F.H. Estupiñán López, C. Gaona Tiburcio, P. Zambrano-Robledo, J.A. Cabral, and F. Almeraya Calderon

Abstract In the aeronautical industry, precipitation hardenable (PH) stainless steels are accessories used in aircraft, turbine blades, and other parts. These alloys are alloys of iron, chromium, and nickel, characterized by mechanical strength obtained through age hardening by heat treatment. The addition of elements such as Al, Ti, Mo, and Cu lead to the appearance of intermetallic compounds. In this work, we performed electrochemical corrosion testing of 17-4 PH and 17-7 PH stainless steels with cyclic polarization and linear polarization resistance tests, showing the susceptibly of these stainless steels to pitting and intergranular corrosion. These tests were carried out in an electrochemical cell with a saturated calomel reference electrode at room temperature. The solutions were hydrochloric acid (HCl), sodium chloride (NaCl), and neutral (H_2O) environments. The results in the neutral environment showed that 17-4 PH had more pitting than 17-7 PH; they had similar behavior in a salty environment. 17-4 PH and 17-7 PH generally presented corrosion in an acid environment. The susceptibility to corrosion of stainless steel types 17-4 PH and 17-7 PH was demonstrated; however, the environment is a vital factor in the behavior of this steel.

Keywords Pitting corrosion • Cyclic polarization • 17-4 and 17-7 PH stainless steels • Linear polarization resistance

---

F.H. Estupiñán López (✉) • C. Gaona Tiburcio • P. Zambrano-Robledo
J.A. Cabral • F. Almeraya Calderon
Universidad Autónoma de Nuevo León, UANL. Facultad de Ingeniería Mecánica y Eléctrica, FIME, Centro de Investigación e Innovación en Ingeniería Aeronáutica, CIIIA, Carretera a Salinas Victoria Km. 2.3 s/n., Apodaca, NL 66600, México
e-mail: festupinan.ciiia@gmail.com

P. Zambrano-Robledo et al. (eds.), *Proceedings of the Symposium of Aeronautical and Aerospace Processes, Materials and Industrial Applications*,
DOI 10.1007/978-3-319-65611-3_6

1 Introduction

In the aeronautical industry, precipitation hardenable (PH) stainless steels are accessories used in aircraft, turbine blades, and other parts. These PH stainless steel alloys contain 15–18% chromium and 4–5% nickel [1], with excellent properties, due to copper precipitates [2]. Stainless steel has high demand in industry because of its inert surface film of CrO, created by contact between oxygen and the chromium in the steel. On the nanoscale this film is distinctive because it is continuous, compact, and adherent [3]. It is a natural anticorrosive (this layer is only about 130 angstroms thick and protects the base stainless steel from corrosion) (Fig. 1). This steel can be martensitic, austenitic, or semiaustenitic. Nowadays, PH steels are a useful family of steels to apply in many components because of their heat treatment characteristics and combination of good corrosion resistance, high strength, low distortion, excellent weldability, and relatively high hardness (up to 49 HRC). These steels are classified by the function of their chemical composition and the different phases present in the microstructure. 17-4 PH stainless steel (AISI type 630 or UNS S17400) is a low-carbon martensitic stainless steel containing nickel and copper, hardenable by precipitation [4]. It receives heat treatment with solubilization at up to 1050 °C and precipitation between 450 and 565 °C. 17-7 PH stainless steel is originally semiaustenitic, also known as AISI-SAE 631 or UNS S17700, which receives heat treatment with solubilization at 1065 °C and precipitation at 760 °C [5].

In this work, 17-4 PH and 17-7 PH stainless steels were selected. These stainless steels are susceptible to pitting corrosion [6]. The corrosion mechanisms and corrosion rates were observed in different environments with electrochemical testing to evaluate the corrosion rates and corrosion behavior, and gravimetric testing in hydrochloric acid (HCl), sodium chloride (NaCl), and neutral (H_2O) environments.

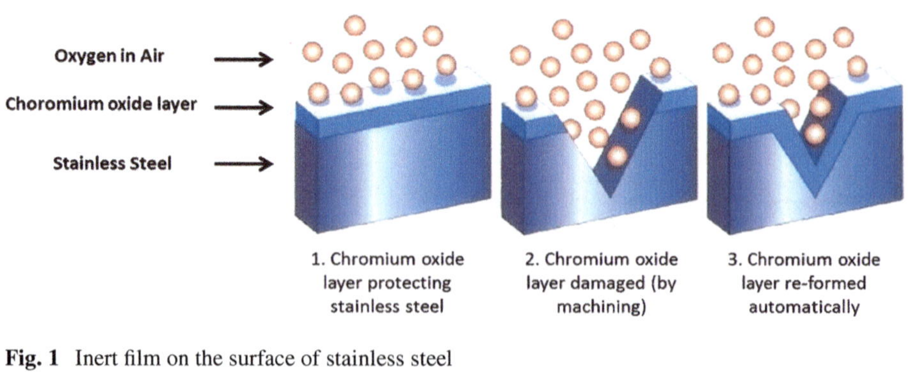

Fig. 1 Inert film on the surface of stainless steel

Fig. 2 Preparation of samples for electrochemical tests

2 Experimental Procedures

The materials used in this study were 17-4 and 17-7 PH stainless steels. Samples were cut from a rolled bar 1 cm in diameter. A copper wire was attached to the samples to guarantee conductivity through the metal. Epoxy resin was used to cover the samples; the exposed area of the specimen was 1 cm^2 (Fig. 2).

The specimens were subjected to a bloom process with emery paper for the electrochemical test. Three specimens of 17-7 PH and 17-4 PH without epoxy resin were bloom processed for the test requirements. Two electrolytes were used for the electrochemical evaluation: for the first, water was used as the base environment; for the second, brine (NaCl 5%) and hydrochloric acid (HCl 5%) were used. Various electrochemical tests allow us to qualitatively assess the tendency of materials to suffer corrosion pitting. In this study the techniques of cyclic potentiodynamic polarization (CCP) [7] and linear polarization resistance (LPR) [8] were carried out in an electrochemical cell with a saturated calomel electrode (SCE) as a reference, platinum as the auxiliary electrode, and the sample as the working electrode (Fig. 3).

The gravimetric testing had a duration of 768 h, where one specimen of the 17-4 PH and 17-7 PH stainless steels were introduced into 50 ml of the three environments above.

3 Results

The microstructures of the materials used in this study are shown in Fig. 4. In Fig. 4a, which shows the microstructure of the 17-4 PH stainless steel, the presence of the martensitic matrix in the solution-annealed condition can be observed. On the other hand, for the 17-7 PH stainless steel (Fig. 4b), the microstructure shows equiaxed grains characteristic of this alloy.

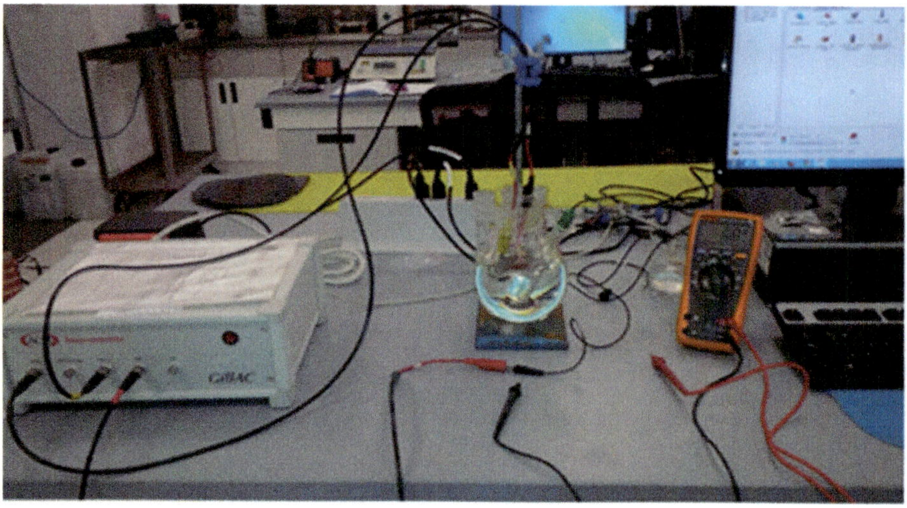

Fig. 3 Cyclic potentiodynamic polarization and linear polarization resistance electrochemical cell

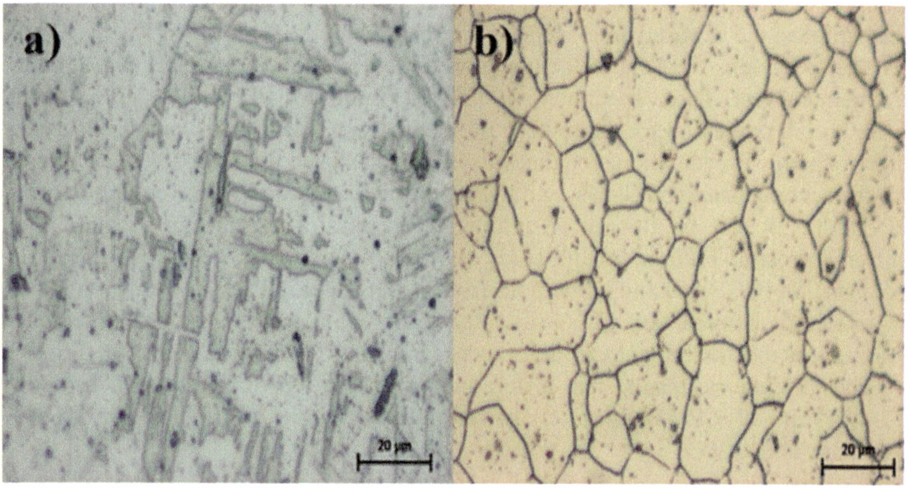

Fig. 4 Microstructures of (**a**) 17-4 PH and (**b**) 17-7 PH stainless steels (magnification 500×)

Figure 5a shows graphs obtained in the three environments of the 17-4 PH and 17-7 PH stainless steels with CCP. The average open circuit potential (OCP) values in H_2O were −20 mV for 17-4 PH and −37 mV for 17-7 PH stainless steels. This environment was not aggressive to the stainless steels [9], which had similar behavior with low current density. In this test, susceptibility to pitting corrosion was observed, due to a hysteresis loop in the reverse scan return inside the anodic loop;

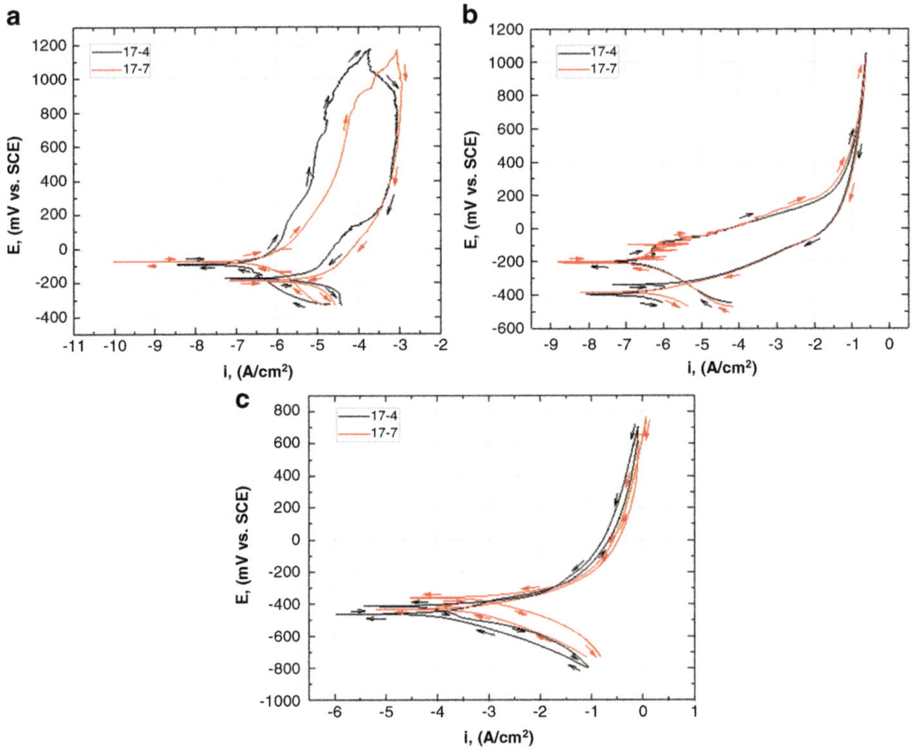

Fig. 5 Cyclic curve polarization of 17-4 PH and 17-7 PH stainless steels in (**a**) H_2O, (**b**) NaCl 5%, and (**c**) HCl 5%

this behavior was indicative of pitting corrosion [10]. The hysteresis amplitude showed the nucleation and growth of the pitting. The stainless steels showed similar behavior in the salt environment, where the average OCP values were −153.9 mV for 17-4 PH and −171.2 mV for 17-7 PH stainless steel, and the current density was slightly higher [8] than in H_2O. In the reverse scan, the hysteresis loop return inside the anodic loop (Fig. 5b) was indicative of pitting corrosion, thought the hysteresis amplitude was much smaller than in H_2O, meaning the pitting was not deep.

The acid environment showed an anodic loop in constant activation, with higher current density and average OCP values of −491.3 mV for 17-4 PH and −476 mV for 17-7 PH stainless steels. The stainless steel behavior showed uniform corrosion, as observed in Fig. 5c, where in a reverse scan, the hysteresis loop returned outside the anodic loop.

Graphs of the linear polarization resistance are shown in Fig. 6. In Fig. 6a it can be observed that in H_2O, the stainless steels' behavior was similar; however, 17-7 PH showed a lesser slope than 17-4 PH stainless steel because its current density was more positive, the current was higher, and the corrosion rate was greater. In the salty environment, 17-4 PH stainless steel showed a lesser slope because it had 3% less

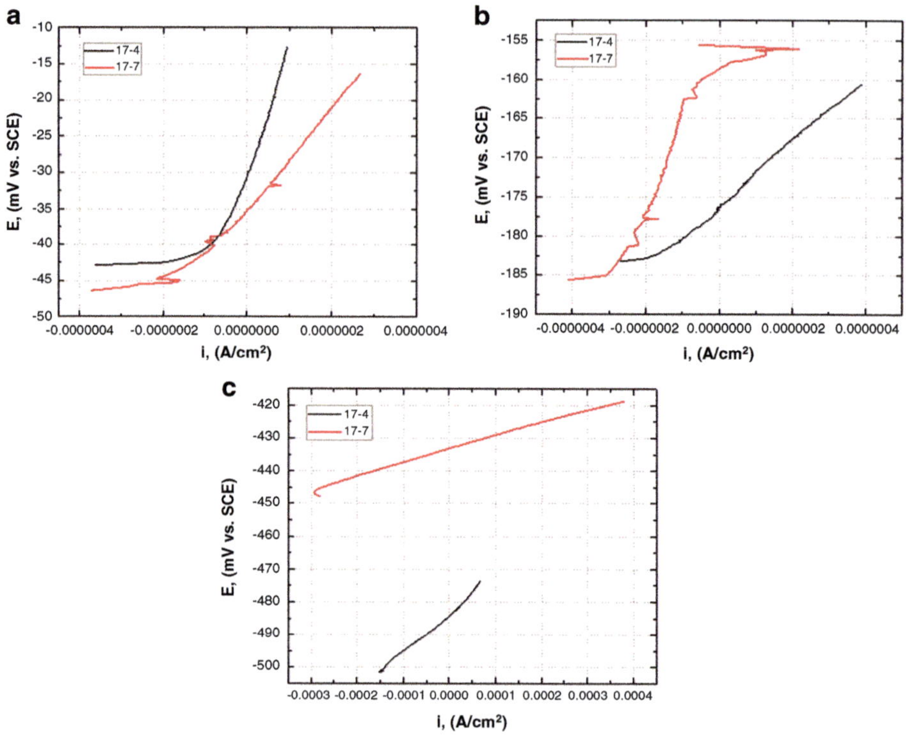

Fig. 6 Linear polarization resistance of 17-4 PH and 17-7 PH stainless steels in (**a**) H_2O, (**b**) NaCl 5%, and (**c**) HCl 5%

nickel than 17-7 PH stainless steel, and the corrosion rate was higher because it had a greater current density than 17-7 PH stainless steel, as can be observed in Fig. 6b. In Fig. 6c, which shows the stainless steels' behavior in an acid environment, a big difference can be seen between 17-4 PH and 17-7 PH stainless steels. While 17-7 PH stainless steel showed more positive values for current density and current potential, 17-4 PH stainless steel showed lesser values; therefore, 17-4 PH stainless steel had a lesser corrosion rate.

Testing for corrosion immersion in the test solutions was performed according to the specifications of ASTM G31 [11]:

- Cleaning of samples (mechanics and chemistry)
- Measurement and weighing of samples (W_0 in grams)
- Immersion of samples in test solutions
- Extraction of samples after 768 h
- Sample cleaning (chemistry)
- Coupon weighing (W_f)

Table 1 Results of gravimetric testing

Stainless steel	Environment	W_0 (g)	W_f (g)	W (g)	Density (g/cm³)	Area (cm²)	Corrosion rate (mmpy)
17-4 PH	HCl	6.0296	5.9374	1.59E-01	7.78	39.153	0.059
	NaCl	5.8162	5.8085	7.50E-03	7.78	38.123	0.0028
	H_2O	6.1656	6.1577	7.60E-03	7.78	40.095	0.0027
17-7 PH	HCl	4.4963	4.334	0.7337	7.81	35.93	0.298
	NaCl	4.7745	4.6871	8.71E-02	7.81	33.823	0.0376
	H_2O	4.4963	4.7876	2.04E-02	7.81	33.4	0.0089

The corrosion rate was calculated using the following equation:

$$\mathrm{CR}\left(\mathrm{mm}/\mathrm{y}\right) = \frac{8.76^{*}10^{4*}\,W}{At\rho} \tag{1}$$

where CR = corrosion rate, W = mass loss in grams, A = exposed area in square centimeters, t = time in hours, and ρ = material density in grams per cubic centimeter.

Table 1 shows the corrosion rates in the gravimetric test.

Figure 7 shows the corrosion rates of the two materials in the three test solutions, obtained from linear polarization resistance and gravimetric tests.

It can be seen that both techniques presented similar results for the corrosion rates; 17-4 PH stainless steel had higher resistance to corrosion in the three electrolytes.

4 Conclusions

- The susceptibility to corrosion of stainless steel types 17-4 PH and 17-7 PH has been demonstrated; however, the environment is a vital factor in the behavior of the steel. Electrochemical tests reveal significant pitting in sodium chloride solution for both materials.
- Sulfuric acid is one of the electrolytes that presents a greater corrosion rate, although its corrosion mechanism in 17-4 PH and 17-7 PH stainless steel is not pitting corrosion. However, 17-7 PH stainless steel has a lower corrosion rate in this electrolyte.
- Both techniques give coherent results for electrochemical parameters.

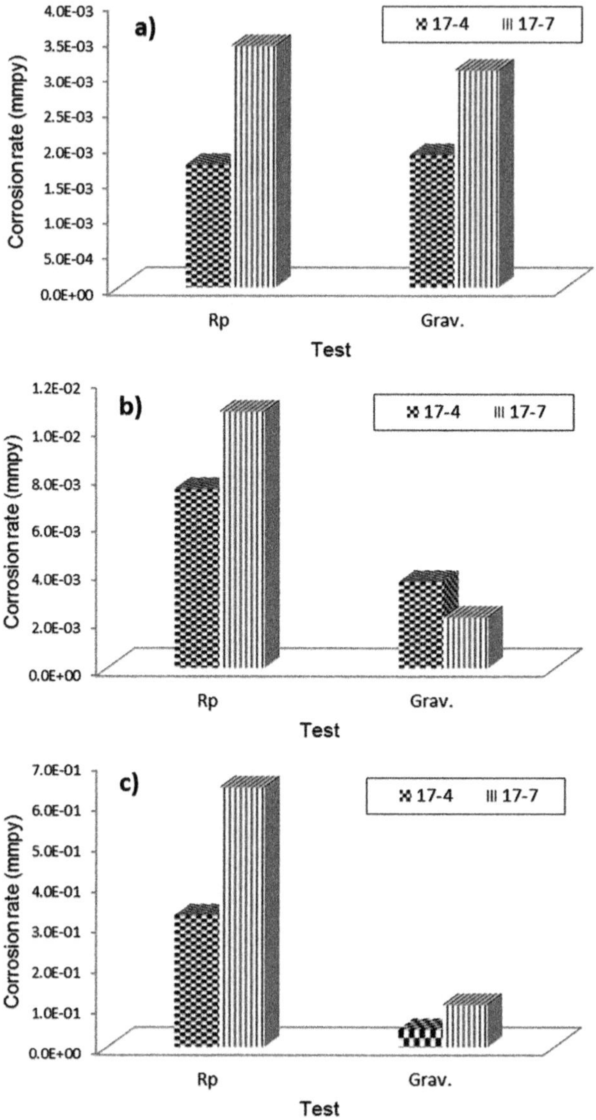

Fig. 7 Corrosion rates obtained from linear polarization resistance and gravimetric tests of 17-4 PH and 17-7 PH stainless steels in (**a**) H_2O, (**b**) NaCl 5%, and (**c**) HCl 5% solutions

Acknowledgments The authors wish to thank the Academic Group UANL-CA-316 of the Autonomous University of Nuevo León and project Promep /103.5/14/10812.

References

1. Solá, P. M. (1990). *Materiales resistentes a la corrosión* (1st ed.). Barcelona: Marcombo. ISBN: 8426707726.
2. Hsiao, C. N., Chiou, C. S., & Yang, J. R. (2002). Aging reactions in a 17-4 PH stainless steel. *Materials Chemistry and Physics, 74*, 134–142.
3. Istrati, G. (1961). *Manual de los aceros inoxidables* (p. 231). Buenos Aires: Alsina.
4. Bressan, J. D., Darosa, D. P., Sokolowski, A., Mesquita, R. A., & Barbosa, C. A. (2008). Influence of hardness on the wear resistance of 17-4 PH stainless steel evaluated by the pin-on-disc testing. *Journal of Materials Processing Technology, 205*, 353–359. doi:10.1016/j.jmatprotec.2007.11.251.
5. Davis, J. R. (2000). *Alloy digest sourcebook: Stainless steel* (pp. 358–359). Materials Park: ASM International.
6. Morral, F. R., Jimeno, E., & Molera, P. (2004). *Metalurgia general* (Vol. 1, tomo 2, p. 1449). Barcelona: Editorial Reverté, S.A.
7. Gaona Tiburcio, C., Estupiñán López, F. H., Zambrano Robledo, P., Cabral M, J. A., Barrios Durtewitz, C., & Almeraya Calderón, F. (2016). Electrochemical corrosion of ferritic 409 and 439 stainless steels 409 and 439 in NaCl and H_2SO_4 solutions. *International Journal of Electrochemical Science, 11*, 1080–1091.
8. Luis Bilurbina Alter, F. L. M., & Laco, J. I. I. (2003). *Corrosión y protección* (Vol. 1, 1st ed., p. 274). Barcelona: Universidad politecnica de Catalunya. SL.
9. Soriano Rull, A., & Pancorbo Floristán, F. J. (2012). *Suministro, distribución y evacuación interior de agua sanitaria* (p. 506). Barcelona: Marcombo, S.A.
10. Evans, U. R. (2003). *Corrosiones metálicas* (p. 318). Sevilla: Editorial Reverté, S.A..
11. ASTM. (2004). *G31-72, standard practice for laboratory immersion corrosion testing of metals*. Philadelphia: American Society for Testing and Materials.

Corrosion Behavior of Austempered Ductile Iron Used in the Aeronautical Industry Evaluated on Acid Solutions

Rodolfo Rocha-Reséndez, J.A. Cabral-Miramontes, C. Gaona-Tiburcio, P. Zambrano-Robledo, F. Estupiñan-López, and F. Almeraya Calderon

Abstract The austempered ductile iron (ADI) is a material which has great potential when talking about applications relating wear resistance, high strength, toughness, and ductility. This material has a unique microstructure, consisting of a matrix of high carbon austenite and a mixture of ferrite and carbides. This microstructure provides mechanical properties much higher than other iron castings. Usually ADI is used on automotive industry, such as suspension parts, arrows, and crankshafts. It is also used on the aeronautical industry, in the building process of aircraft landing gear. In certain applications where ADI is used, its corrosion resistance must be a transcendental property; nonetheless owing to its structure, the ADI doesn't possess pretty noticeable resistance to corrosion. Despite this, it's of huge interest to know how different parameters on heat treatment (applied to ADI) may affect the ADI microstructure, and it can disturb its behavior due to those variables, and in consequence its application areas could be reduced. In this paper we analyze the corrosion resistance presented by ADI irons, after having been subjected to different austempered treatments (different temperatures within the austenitizing muffle furnace and salt bath). The austenitizing temperatures evaluated were 815, 870, and 930 °C, where the waiting time used was 90 min. On the other hand, the salt bath was used at a temperature of 250 and 400 °C, having the waiting time of 60 min. The corrosion resistance was evaluated on two different solutions (NaCl and H_2SO_4), supporting this with two different electrochemical techniques, the linear polarization resistance (LPR) and potentiodynamic polarization curves (CPC). The outcomes show that the temperatures of the salt bath play a crucial role in evaluating the corrosion resistance of these irons.

Keywords ADI • Heat treatment • Corrosion

R. Rocha-Reséndez (✉) • J.A. Cabral-Miramontes • C. Gaona-Tiburcio
P. Zambrano-Robledo • F. Estupiñan-López • F. Almeraya Calderon
Universidad Autónoma de Nuevo León, Facultad de Ingeniería Mecánica y Eléctrica, San Nicolás de los Garza, NL, Mexico

Centro de Investigación e Innovación en Ingeniería Aeronáutica (CIIIA), Apodaca, NL, Mexico
e-mail: rodolfoo.rochaa@gmail.com

P. Zambrano-Robledo et al. (eds.), *Proceedings of the Symposium of Aeronautical and Aerospace Processes, Materials and Industrial Applications*, DOI 10.1007/978-3-319-65611-3_7

1 Introduction

Nodular iron is one of the most actively developing cast materials in the world; this is the favorite material by engineers for applications involving parts of machinery and equipment. Because of this, there is a necessity to improve its functional properties by means of development of new kinds of ductile (nodular) irons in order to satisfy the requirements for each type of application or properties. Some of that developments have produced different types of ductile iron (austenitic ductile iron, austempered ductile iron, martempered ductile iron) [1]; in our case study we will analyze the austempered ductile iron.

In fresh years the austempered ductile iron (ADI) has been engaged much attention by cause of its notable combination of hardness, strength, ductility, and impact resistance; these attractive properties are related to the unique ADI's microstructure known as ausferrite, which is a mixture of bainitic ferrite and high carbon austenite [2].

The ADI is an alloyed ductile iron that has been submitted into a two-step heat treatment process known as austempering. In the first step the casting is initially heated to the austenitizing temperature range and above for sufficient time in order to obtain a fully austenitic matrix saturated with carbon. After austenitizing, the alloy is austempered (second step); quenched rapidly to a temperature range amid 250 and 400 °C inside of a molten salt bath and kept there during 2–4 h [3, 4]; after holding for a preselected time at the austempering temperature, the alloy is cooled to room temperature by either water quenching or air cooling [5].

During austempering heat treatment, the fully austenitic matrix transforms into acicular ferrite and high carbon austenite; this is a matrix called ausferrite [3]. Suitable austempering process produces a particular bainitic structure (named ausferrite) which consists of high carbon and acicular ferrite with graphite nodules dispersed in it; the nature of this bainitic microstructure depends on the austempering temperature and time [4]. There are two isothermal temperature ranges to produce ADI; low-temperature ADIs (temperature 250–330 °C) have heightened hardness, high strength, and low ductility. Moreover the strength decreases with the rising temperature of isothermal transformation (350–400 °C), while the ductility increases together [6]. The control of the austempering temperature is the most adequate technique to control the austempered microstructure and properties [5]. On the other hand, the austempering time ought to be hence long enough to guarantee that high carbon austenite is obtained, and the development of martensite is avoided. If this time is too long, nevertheless, the high carbon content austenite becomes saturated with carbon and dissolves into ferrite and carbide [3].

The remarkable microstructure features are the presence or absence of carbides in austenite or ferrite, morphology of ferrite, carbon content of austenite, and retained austenite content [7]. A proper heat treatment will avoid the formation of unwanted microstructural constituents such as martensite or pearlite [4]. Regardless of that, with difference of time and temperature of austenitizing and austempering, it is possible to achieve different combinations of properties that can be specifically designed and adjusted for a certain purpose [8].

It has been demonstrated that ADI austempered at higher temperature showed better corrosion resistance than the ones austempered at lower temperature; the sophisticated nature of corrosion of ADI is influenced by contents of ferrite, retained austenite, and the carbon content in a collective way [9]; as important fact it was reported that the percentage of retained austenite in ADI influences its corrosion behavior; the higher the percentage of retained austenite, the lower the corrosion rate [10].

Since ADI was discovered, many modifications in technology of heat treatment leading to cost saving and/or property improvement were proposed, for example, direct austempered ductile iron (DADI) and ausformed-austempered ductile iron (AADI) [11]. Besides, the manufacturing cost of ADI (heat treatment and casting) is also substantially lower, for example, against the cost of carburizing or forged steel (increasing the cost of operation by the plastic forming) [1, 4]. Additionally as a consequence of the unique properties of ADI such as high strength, fatigue behavior, and ductility, ADI has been considered a very up-and-coming engineering material and is an economical replacement for forged steel or even aluminum alloys in several structural applications in the automotive industry (such as transmission gears, crankshafts, connecting rods), defense industry (aircraft landing gears, cannon shells), and railroad industry [4].

It is important to study the corrosion behavior of this material while it is exposed into acid or chlorine media, obviously knowing that automotive and aircraft parts which are made by ADI are in continuous contact with this kind of media [12]. The austempered ductile iron (ADI) present different corrosion behaviors while being evaluated in H_2SO_4 compared to aqueous solitions containing Cl^- ions [13].

From the above, in this work we evaluated the corrosion behavior presented by ADI in solutions of NaCl and H_2SO_4, having different austenitizing temperatures and different austempering temperatures at immersion of molten salts. This analysis was evaluated using two different electrochemical techniques as linear polarization resistance [14] and potentiodynamic polarization curves [15]. The good mechanical properties provided by the ADI, having good corrosion resistance, involve a bigger scope to this material.

2 Experimental Procedure

2.1 Material

To obtain the ductile iron and then austempering ductile iron with pearlite microstructure, the following chemical compositions were used for the experiments: C 3.66%, Mn 0.42%, Si 2.77%, Ni 0.42%, Cr 0.048%, Mo 0.046%, and Cu 0.47%.

Table 1 Heat treatments employed to manufacture the ADIs

Conventional muffle furnace (austenitizing)		Salt bath oven (austempering)		
Time (min)	Temperature (°C)	Temperature (°C)	Time (min)	Nomenclature
90	815	250	60	815-1
		400		815-2
	870	250		870-1
		400		870-2
	930	250		930-1
		400		930-2

2.2 Heat Treatments

As showed in the bibliography, the austempered process consists of two steps [4, 5]; first the austenitizing treatment was realized with the aid of a conventional muffle furnace, and second in the austempering process, salt bath was used. All the variables in the preparation of ADI, referring to different temperatures (in both heat treatments) handled, are shown in Table 1.

2.3 Microstructure

All evaluated pieces were prepared by metallographic techniques, first by grinding with sandpapers of SiC, then with polishing to mirror finish. Afterward the specimens were chemically attacked with nital 2% to determine the microstructure of all samples obtained, all previous by using optical microscopy technique.

2.4 Corrosion Characterization

The electrochemical techniques were carried out with a potentiostat/galvanostat/ZRA Gill-1 AC. The techniques applied were linear polarization resistance (LPR) standard ASTM G59 [14], with operation parameters of ±20 mV with respect to corrosion potential (Ecorr) and with scanning speed of 10 mV/min in order to determine the corrosion speed of ADI. On the other hand the potentiodynamic polarization curves (PPC) were begun with a potential of -300 mV with respect to corrosion potential and with 60 mV/min as scanning speed and final potential of 1200 mV, all this according to the standard ASTM G5 [15].

All tests were performed with the help of a coating evaluation cell, by using a design of three electrodes: one reference electrode of saturated calomel (SCE), auxiliary electrode of platinum, and work electrode of the different ADIs obtained by

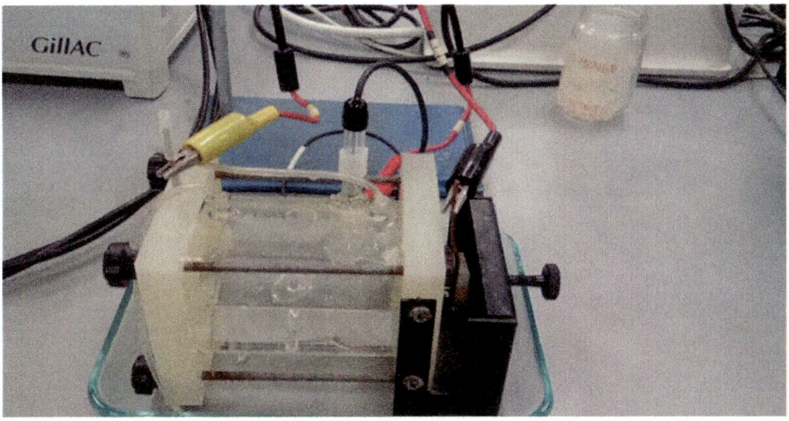

Fig. 1 Electrochemical cell employed in this work. #1 Auxiliary electrode of platinum, #2 Reference electrode and #3 Work electrode

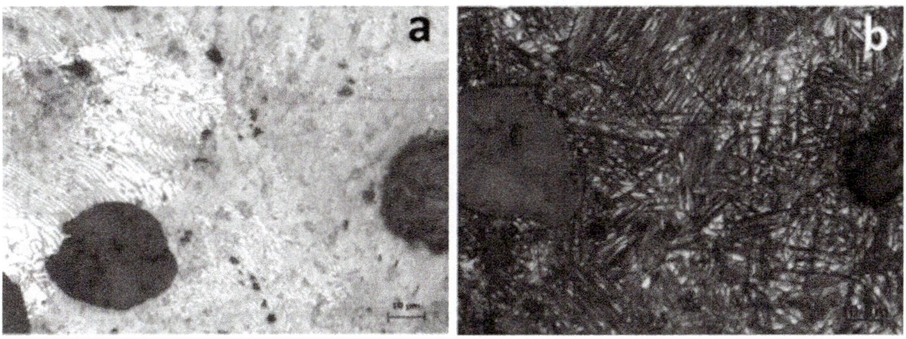

Fig. 2 Microstructure of nodular iron (**a**) and ADI (**b**)

different heat treatments. This electrochemical arrangement is showed in Fig. 1. Each ADI specimen was tested in electrolytes of sulfuric acid 0.5 N (H_2SO_4) and sodium chloride at 3.5% of concentration (NaCl); all the assessments were practiced in room temperature.

3 Results and Discussion

3.1 ADI's Microstructure

With the characterization used, it could watch the different phases at the microstructure of ADI. In Fig. 2 we observe the microstructure of ductile iron and ADI (chemically attacked with nital 2%); at first image we watch that the microstructure of nodular iron is constituted mainly by ferrite and pearlite, with nodule's presence. On the other hand ADI microstructure is chiefly formed by ausferrite and nodules.

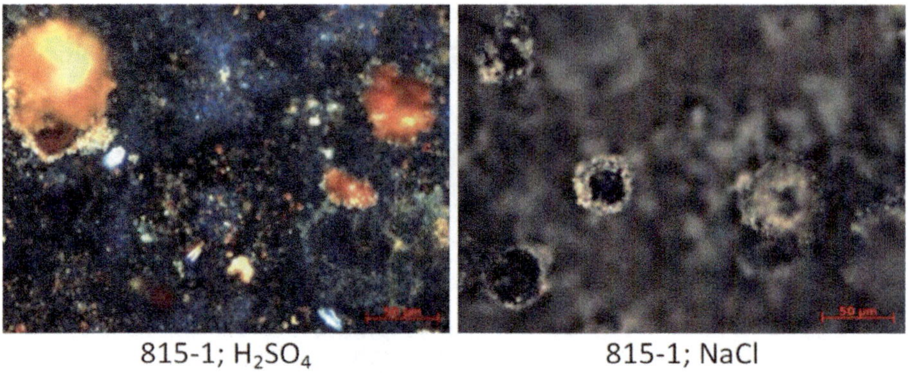

Fig. 3 Microstructure of ADI attacked with the electrochemical mediums at 815-1 conditions used (20×)

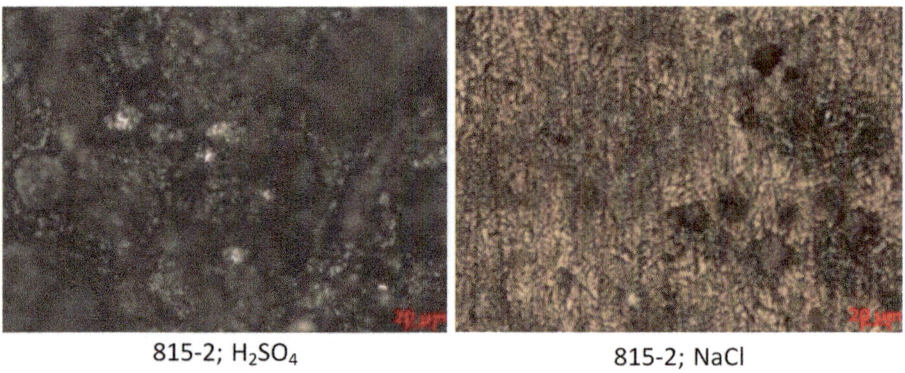

Fig. 4 Microstructure of ADI attacked with the electrochemical mediums at 815-2 conditions used (20×)

After performed the respective heat treatments with their correspondent electrochemical test, the ADI's microstructure was analyzed anew in order to look the behavior or answer of this microstructure against the different electrochemical mediums. In Figs. 3 and 4, the microstructure of different samples (with different variables of heat treatment) evaluated in the electrochemical mediums (NaCl and H_2SO_4) is presented.

It is observed how all metal surfaces present corrosion except the nodules which act like cathodic areas of ADI.

3.2 *Linear Polarization Resistance*

Once the LPR test was performed to every ADIs, the results obtained showed different corrosion rates (in each solution evaluated).

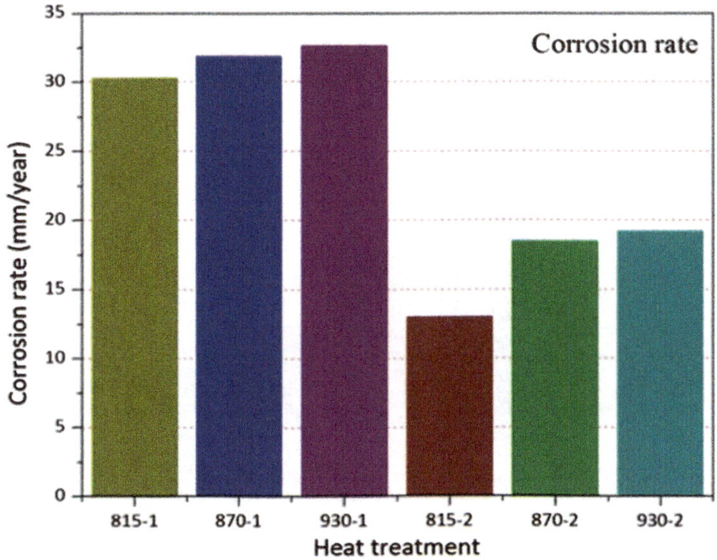

Graph 1 Corrosion rate evaluated in H_2SO_4

In Graph 1 the different results obtained for corrosion rate of ADIs evaluated in H_2SO_4 solution are presented. It is observed that in heat treatment which involves an austempering temperature of 250 °C, the corrosion rate reaches 30 mm/year, on the other hand, for samples evaluated on 400 °C, the corrosion rate presented was minor, reaching 17 mm/year, which is a big difference because we are talking that expands the utility of the material. So according to the bibliography, the corrosion rate of all ADIs will decrease by increasing the austempering temperature [10].

Also it can be observed that the austenitizing temperature affects in the same way the corrosion resistance, inasmuch as the specimens evaluated with higher temperatures showed major corrosion rate than those evaluated in the lower austenitizing temperature. It can be see, that as long as the austempering temperature increases and the austenitizing temperautre decreases, the corrosion rate will be lower.

In Graph 2 the results of corrosion rate of ADI using the same test are shown but now with the ADIs evaluated in NaCl. It can be observed that again there is a change in corrosion rate; anew the specimens treated with 250 °C in the salt bath present an average corrosion rate of 3 mm/year, on the other hand, the samples treated with 400 °C showed minor corrosion rate, with an average of 1.8 mm/. Again it is presented that with increasing austempering temperature there will be an increase in the useful life of ADIs.

As additional information, the ADI presented remarkable corrosion resistance against the saline medium compared with the acid medium. So we could say that ADI will have longer useful life in saline medium than in acid medium.

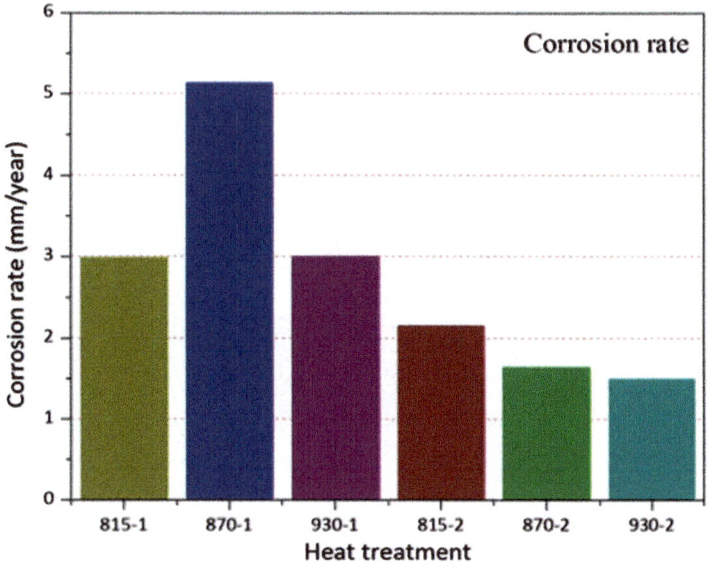

Graph 2 Corrosion rate evaluated in NaCl

3.3 Cyclic Polarization Curves

The assessments of cyclic polarization curves were performed to all ADI evaluated with different electrochemical solutions.

We appreciate in Graph 3 the results of the test evaluated in H_2SO_4. We observe that all graphs present current demand in the anodic zone of the curve of ADIs evaluated with 250 and 400 °C in the salt bath, confirming again the results obtained in LPR test. In Graph 4 the results of cyclic polarization curves performed to ADIs using the solution of NaCl are presented; we observed again that all materials are in continuous activation, without presenting active-passive transitions in none of them. In the same way, the ADIs with the same temperature in the salt bath present similar behavior.

4 Conclusions

- Parameters of austempering heat treatment will affect the corrosion resistance of the ADIs.
- Samples evaluated on 400 °C as austempering temperature presented better corrosion resistance against the samples evaluated on 250 °C.

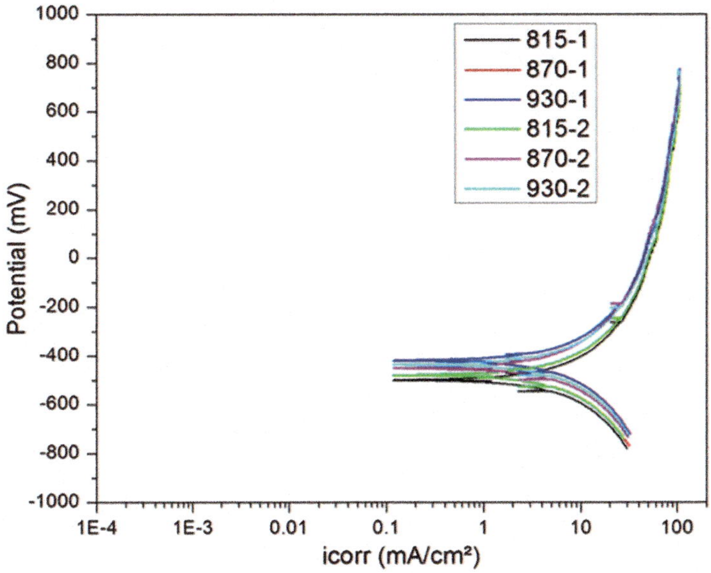

Graph 3 Polarization curves of ADI evaluated in H_2SO_4

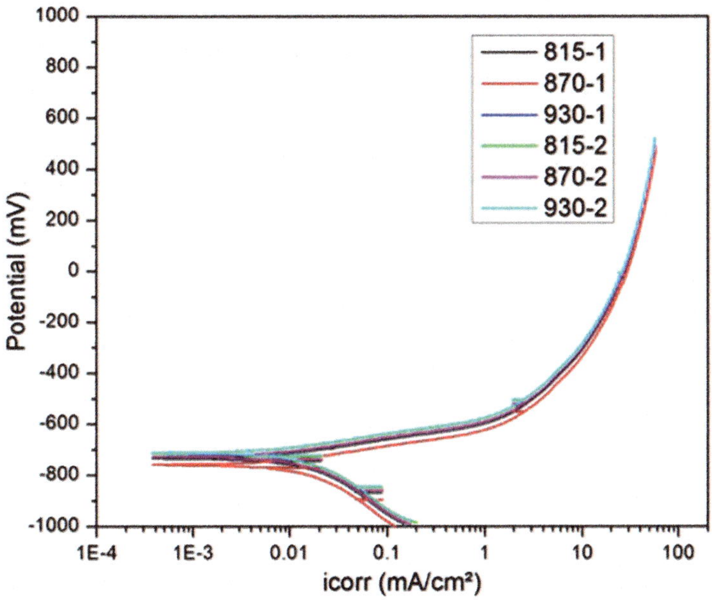

Graph 4 Polarization curves of ADI evaluated in NaCl

- The lowest austenitizing temperature resulted in better corrosion resistance in comparison with the other (higher) austenitizing temperatures.
- ADIs showed major corrosion resistance immersed in the NaCl media in comparison to H_2SO_4.

Acknowledgments To CONACYT for the financial support granted for the development of this research and to the academic group UANL-CA-316 for the facilities granted.

References

1. Wilk-Kolodziejczyk, D., Regulski, K., & Gumienny, G. (2016). Comparative analysis of the properties of the nodular cast iron with carbides and the austempered ductile iron with use of the machine learning and the support vector machine. *The International Journal of Advanced Manufacturing Technology, 87*, 1077–1093. doi:10.1007/s00170-016-8510-y.
2. Guo, X. (2014). Correlation between austempering parameters and hardness of austempered ductile iron based on artificial neural network. *Computer Modeling & New Technologies, 18*, 72–76.
3. Olofsson, J., Larsson, D., & Svensson, I. (2011). Effect of austempering on plastic behavior of some austempered ductile iron alloys. *Metallurgical and Materials Transactions A, 42*, 3999–4007. doi:10.1007/s11661-011-0796-7.
4. Putatunda, S. (2001). Development of austempered ductile cast iron (ADI) with simultaneous high yield strength and fracture toughness by a novel two-step austempering process. *Materials Science & Engineering, 315*, 70–80.
5. Batra, U. (2005). Fracture behavior and mechanism in austempered ductile iron. *Journal of Failure Analysis and Prevention, 5*, 75–81. doi:10.1361/154770205X65936.
6. Janjić, M., Avdušinović, H., Jurković, Z., Bikić, F., & Savićević, S. (2016). Influence of austempering heat treatment on mechanical and corrosion properties of ductile iron samples. *Metabk, 55*, 325–328.
7. Putatunda, S., & Gadicherla, P. (2000). Effect of AustemperingTime on mechanical properties of a low manganese austempered ductile iron. *Journal of Materials Engineering and Performance, 9*, 193–203. doi:10.1361/105994900770346150.
8. Šolić, S., Godec, M., Schauperl, Z., & Donik, Č. (2016). Improvement in abrasion wear resistance and microstructural changes with deep cryogenic treatment of austempered ductile cast iron (ADI). *Metallurgical and Materials Transactions A, 47*, 5058–5070. doi:10.1007/s11661-016-3659-4.
9. Pooja, M., Vijeesh, V., Surendranathan, A., Udupa, K., & Samuel, K. (2017). Corrosion behavior of austempered ductile iron (ADI) in iron ore slurry. *International Journal of Engineering, Science and Technology, 8*, 7–12.
10. Hsu, C., & Chen, M. (2010). Corrosion behavior of nickel alloyed and austempered ductile irons in 3.5% sodium chloride. *Corrosion Science, 52*, 2945–2949.
11. Krzyńska, A., & Kochański, A. (2014). Austenitization of Ferritic Ductile Iron. *Archives of Foundry Engineering, 14*, 49–54.
12. Goyos Pérez, L., Castañeda Segarra, R., Wettinck, E., Valera, R., & Moors, M. (1999). Comportamiento del hierro nodular austemperizado en condiciones de corrosión y desgaste. *Ingeniería Mecánica, 2*, 7–13.
13. Afolabi, A. (2011). Effect of austempering temperature and time on corrosion behaviour of ductile iron in chloride and acidic media. *Anti-Corrosion Methods and Materials, 58*, 190–195. doi:10.1108/00035591111148902.
14. ASTM G59. (1994). Standard reference method for conducting potentiodynamic polarization resistance measurement, ASTM annual book of standards, Vol. 02,03.
15. ASTM G59. (1994). Standard reference test method for making potentiostatic and potentiodynamic anode polarization measurement, ASTM annual book of standards, Vol. 02,03.

Design of Wind Tunnel for Testing Stress Corrosion Cracking

B. Bermúdez- Reyes, P. Zambrano-Robledo, F. Almeraya Calderon, C. Zárate-Tomás, A. Medrano-Mejía, and R. Vargas-Bernal

Abstract Corrosion is one of the greatest threats to civil and military aircraft. In general, the materials that make up aircraft structures are subjected to stresses, causing embrittlement and consequent cracks and fractures. Corrosion and erosion of the aircraft coating, induced by the dynamic chemical environment of the atmosphere, manifests as pitting on the surface material. Because they concentrate the stress, these pits are vulnerable to stress corrosion cracking. In current stress corrosion cracking tests on aeronautics materials, parts constructed from corrosive media are held in universal machines. However, the obtained results do not reflect the flight conditions of the aircraft. Therefore, we propose wind tunnel tests that can assess stress corrosion cracking under simulated flight conditions.

Keywords Stress corrosion cracking • Wind tunnel • Aeronautics materials

---

B. Bermúdez- Reyes (✉) • P. Zambrano-Robledo • F. Almeraya Calderon
Universidad Autónoma de Nuevo León, Centro de Investigación e Innovación en Ingeniería Aeronáutica, Facultad de Ingeniería Mecánica y Eléctrica, Carretera A Salinas Victoria, km 2.3, C. P. 66600 Apodaca, NL, Mexico
e-mail: barbara.bermudezry@uanl.edu.mx; Patricia.zambranor@uanl.edu.mx; falmeraya.uanl.ciiia@gmail.com

C. Zárate-Tomás
Industria Militar Aeronáutica, Avenida Industria Militar 1111, Lomas de Tecamachalco, C. P. 53050 Naucalpan de Juárez, Estado de México, Mexico
e-mail: ciro.zarate.tomas@gmail.com

A. Medrano-Mejía
Laboratorio de Aerodinámica, Escuela Superior de Ingeniería Mecánica y Eléctrica Unidad Profesional Ticomán del Instituto Politécnico Nacional, Av. Ticomán No. 600, Col. San José Ticomán, C.P. 07340 Delegación Gustavo A. Madero, Ciudad de México, Mexico
e-mail: excalibur0937@yahoo.com.mx

R. Vargas-Bernal
Departamento de Ingeniería en Materiales, Instituto Tecnológico Superior de Irapuato, Carretera Irapuato-Silao Km. 12.5, Col. El Copal, C.P. 36821 Irapuato, Guanajuato, Mexico
e-mail: ravargas@itesi.edu.mx

P. Zambrano-Robledo et al. (eds.), *Proceedings of the Symposium of Aeronautical and Aerospace Processes, Materials and Industrial Applications*,
DOI 10.1007/978-3-319-65611-3_8

1 Introduction

Flying aircraft are exposed to atmospheric environments (saline, acidic, and basic) and suspended particles, which corrode the aircraft materials. Additionally, they are subjected to mechanical stresses (such as tension, shear, flexion, torsion, etc.) during takeoff, flight, and landing [1]. The simultaneous effects of corrosive environments and mechanical stress lead to stress corrosion cracking (SCC). Over time, these conditions induce the progressive formation of microcracks that accelerate fatigue [2]. In the design phase of the aircraft, corrosion is inspected only a few times, but during operations the corrosion inspections are exhaustive because corrosion causes structural damage that compromises the structural integrity [3]. The consequent costs to the aeronautics industry are very high, and techniques for predicting corrosion by traditional modeling (the electrochemical mechanism of corrosion) and survival analysis (based on statistical data) have been developed around the world [3].

Aircraft and aeronautics devices are exposed not only to structural damage from SCC but also to stress concentration around holes, inclusions, and pit corrosion on the surface, which favors ductile cracking. The residual stress imposed by the manufacturing processes (such as union processing and machinery) must also be considered [4]. Some SCC researches have attempted to simulate the environments that expose all materials to residual stress, such as high and low temperatures and in-flight damage conditions [5]. Other models detect fatigue and exfoliation on the wind skin [6, 7]. All of these corrosion and fatigue models of aircraft structures are based on a life management framework, which determines the lifecycle of the craft under different environmental effects. The test process employs different corrosion models and methodologies to determinate the exfoliation, pitting, and fatigue [3].

Wind engineering began with the development of wind tunnels in 1871. Since that time, wind engineering has considerably advanced, and wind tunnels are used in the testing of buildings, towers, and airplanes under wind actions [8]. Cermak established that wind tunnels should be designed appropriately for applications, characteristics, facilities, and tests [9]. The airflow in wind tunnels experiences energy loss due to the roughness of walls or friction injection into the fluid, internal turbulence, and sudden changes of transversal areas imposed by limit layers and imperfect edges [10].

In the present work, we design a wind tunnel for testing SCC under flight conditions.

2 Methodology

2.1 Tunnel Design

The wind tunnel for investigating SCC adopted an open circuit design and a centrifugal blower that protects the fan from the corrosive media used in the test.

Table 1 Component and condition data

Data	Values
Fan exit transversal section	1.26×0.66 m
Transversal section (A_i)	0.16 m^2
Test section velocity (V_s)	55.55 m/s
Monterrey's air density (ρ)	1.19 kg/m^3
Mass flow ($\dot{m}$)	28 m^3/s
Monterrey's altitude (h)	537 msl
Suction	0
Static drop pressure (K)	8 cm of water

The design included the following components:

- Engine power calculus
- Wide-angle diffuser design
- Contraction design

To calculate the parameters in each of these components, we require the data listed in Table 1. It is necessary to mention that these data were calculated previously.

The design analysis and planes were constructed in the multi-platform software package CATIA V5, and the calculated data were validated in the preliminary dimension 21 program developed by Universidad Politecnica de Madrid (UPM) [11]. The flow analysis was performed in ANSYS.

2.2 *Preliminary Test*

The preliminary test was conducted on an airfoil. The airfoil profile was coated with kaolin dispersed in acetone. Because the acetone evaporates rapidly, it served only as a vehicle for spreading the kaolin. The coated profile was tested in a wind tunnel with turbulent flow at 30 m/s.

3 Results and Discussion

3.1 *Stress Corrosion Tunnel Design*

The first step was calculating the Reynolds number. In this case, the flow was turbulent ($Re = 3 \times 10^6$) because laminar flows do not exist in real flights, so the suspended particles follow random trajectories [12]. Consequently, the particles

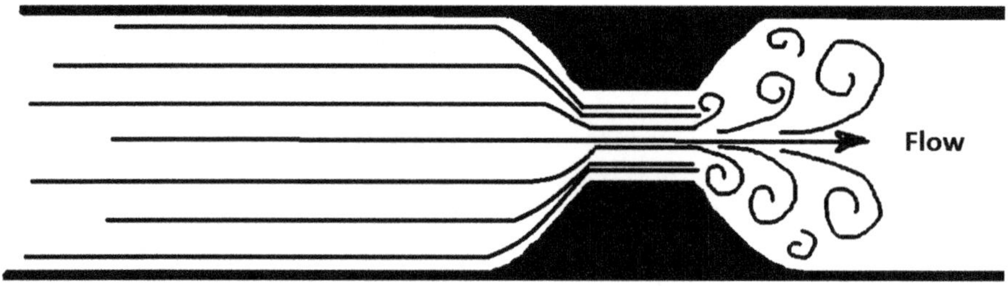

Fig. 1 Airflow through the contraction area

strike all surfaces of the aircraft, provoking microimpacts that damage the structure. As mentioned in Anderson's text, at $Re > 3 \times 10^6$ the limit layer transits directly to turbulent flow (see Fig. 1) [13].

For designing a wind tunnel that assesses SCC under flight conditions, we require the following parameters:

- Engine power
- Wide-angle diffuser angle
- Contraction

Once the relevant data are collected, we can design the tunnel.

Engine Power The power of the engine (P_t) is given by [14]:

$$P_t = 0.5 A_i V_s^3 \tag{1}$$

The value obtained was 32.06 KW; therefore the motor power must be at least 37.28 KW to generate a wind velocity of 55.55 m/s.

Wide-Angle Diffuser Design The diffuser decreases the flow to minimize (as far as possible) the energy loss, that is, to maximize (as far as possible) the recovery pressure [15]. The flow through a diffuser depends on the diffuser geometry, which is defined by the diffuser angle in the cross section [16].

The important parameters in a wide-angle diffuser are the transversal area of the diffuser exit A_s, the diffuser angle (2θ), the pressure drop coefficient (K), and the number of screens in the diffuser (n). Specifically, we must determine the relation between the transversal areas shaped by the diffuser and test section, the diffuser angle 2θ, and the pressure coefficient K in each screen.

2θ is the angle formed by the two lines that join the entry and exit points in the cross section (see Fig. 2). Mehta recommends $2\theta > 6°$ for this diffuser type [16].

Figure 3 shows the area–angle relationships of the diffuser. In this work, the screen number is zero, and the possible 2θ angles lie along the CD line, which corresponds to 0 screen line or Kline line [16].

According to the graph, $2\theta = 17°$ and the squareness of the diffuser exit is calculated as [17]:

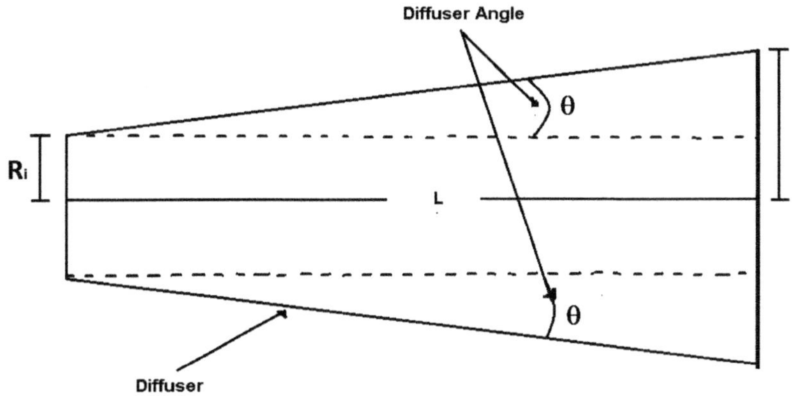

Fig. 2 Illustration of angle 2θ in the diffuser cross section

$$N = \frac{A_e}{A_i} \tag{2}$$

Because the boundary layer does not separate in the turbulent regime, $N = 3$ in this case. The lack of boundary layer separation was reported by Chmielewski, who conducted numerical simulations of turbulent flows with $Re = 10^6$ [18]. According to Mikhail, the squareness of the distribution shape refers to the square geometry of the diffuser transversal area from the contraction chamber to the diffuser exit [16]. From the previous data, A_e was calculated as 2.52 m^2. The geometrical meanings of A_e and A_i are illustrated in Fig. 4.

The total diffuser length was calculated from the general expression of the duct geometry [19]:

$$\theta_s = \tan^{-1}\left(\frac{R_s - R_i}{L}\right) \tag{3}$$

where:

R_s = exit hydraulic radius
R_i = inlet hydraulic radius
L = Duct length

The total length is 1.234 m. We can then calculate the cone contraction that accelerates the flow from the setting chamber to the test section. Bell and Mehta designed the cone contraction by the following fifth-order polynomial function [20, 21]:

$$h = H_i - [H_i - H_e]\left[6(\xi)^5 - 15(\xi)^4 + 10(\xi)^3\right] \tag{4}$$

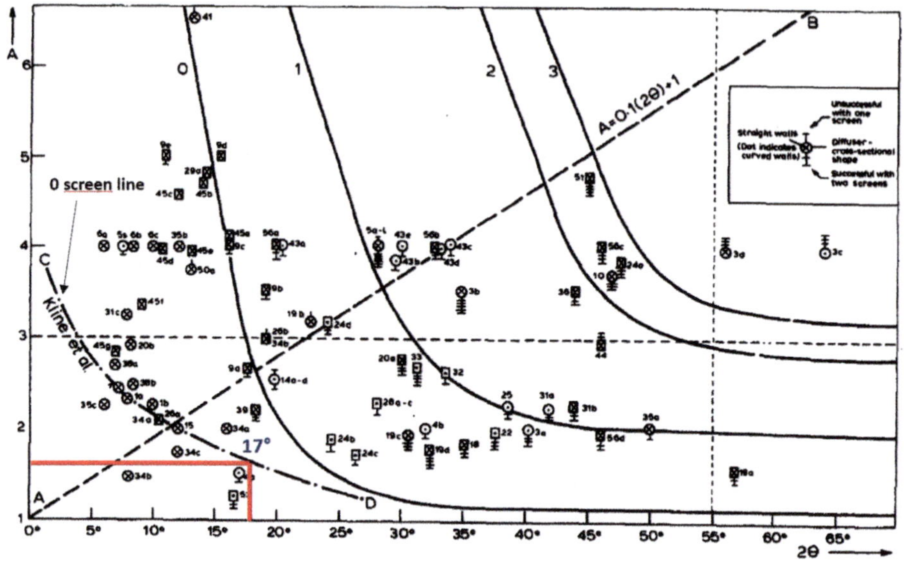

Fig. 3 Area (A) versus diffuser angle (2θ) relationships for the diffuser type shown in Fig. 1 [16]

Fig. 4 Inlet and exit of the diffuser

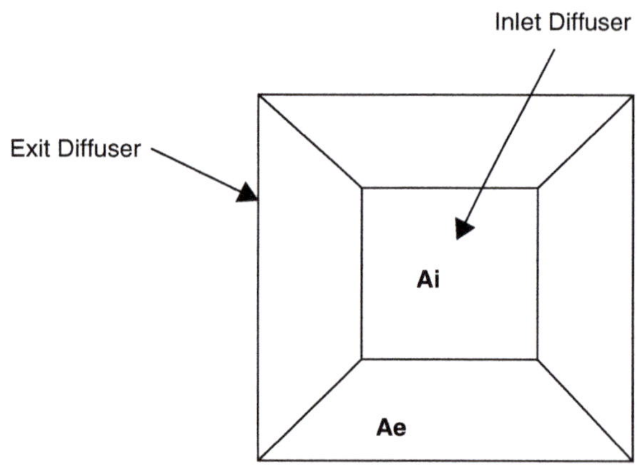

where:

$\xi = \dfrac{X}{L}$ is the normalized length
H_e = height of contraction wall from the symmetry axis of the exit
H_i = height of contraction wall from the symmetry axis of the inlet

To use this polynomial, we require a starting point (X_n, Y_n), a midpoint (X_m, Y_m), and a final point (X_f, Y_f), as shown in Fig. 5. In the present case, we have:

(X_n, Y_n) = (0, 0).
(X_m, Y_m) = (0.67, 0.39).
(X_f, Y_f): = (1.234, 0.39).

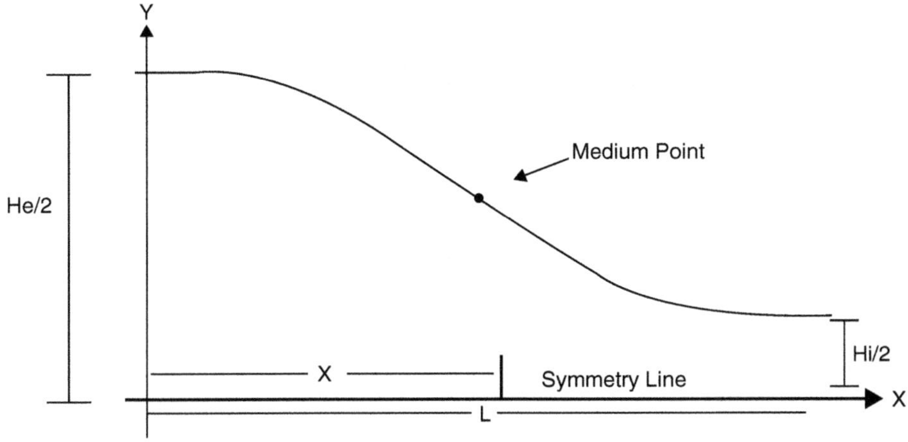

Fig. 5 Bell–Mehta fifth-order polynomial function, used to design the cone contraction

To verify the calculus, all data were corroborated by the software dimension 21 (Fig. 6).

Figure 5 describes the characteristic fifth-order polynomial curve of Bell and Mehta and the geometrical parameters of the contraction wall profiles between the inlet and exit. The curves obtained by a mean bipartition numerical method and preliminary dimension 21 software are very similar (see Fig. 6). Therefore, the profile of the wind tunnel has been correctly designed for SCC assessment.

Cattafesta and Nordin mentioned the critical role of inlet contraction in determining the flow in the test section. Moreover, the shape of the contraction dictates the turbulence intensity level in this section [21, 22].

After performing the calculus, the final structural concept was built in CATIA (Fig. 7).

The flow velocities into the various sections of this wind tunnel configuration were simulated in ANSYS (Fig. 8). The color scale indicates the velocities in each stage of the tunnel. The maximum speed (55.55 m/s) was reached in the test section. At this velocity, the particles will be optimally accelerated in the flow and will impact on the aero parts (aerodynamic profiles, fuselages, empennages, turbine blades, etc.), generating SCC under flight conditions.

3.2 *Preliminary Test*

Panels (a) and (b) of Fig. 9 show the aerodynamic profiles before and after the test, respectively. The airfoil surface is initially homogeneous. After the test, parts of the surface remain uniform, but large areas have been stripped of their kaolin powder, and some random streamlines are visible. Moreover, the kaolin accumulated on the surface at the 1/3 chord line, near the pressure center. When approaching some areas of the profile, the turbulent flow becomes perpendicular to the wind flow. As shown

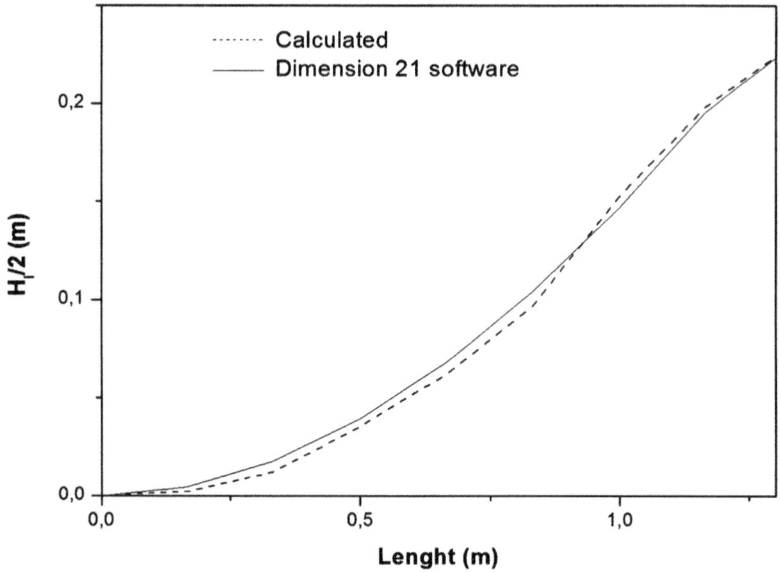

Fig. 6 Bell–Mehta fifth-order polynomial function for designing the cone contraction of stress corrosion cracking in a wind tunnel, obtained by a numerical method (*dotted curve*) and by dimension 21 software (*solid curve*)

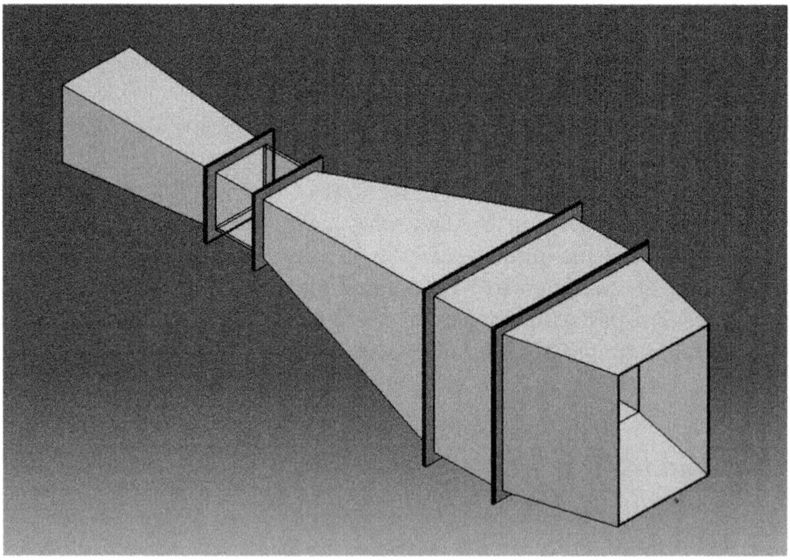

Fig. 7 Simulated structure of wind tunnel for assessing stress corrosion cracking

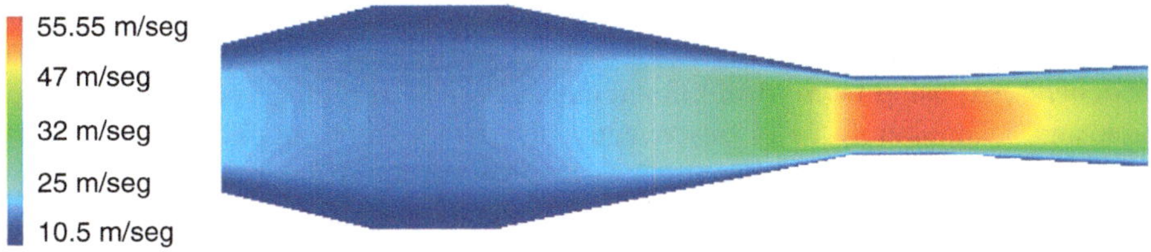

Fig. 8 Velocity scale in the designed SCC wind tunnel

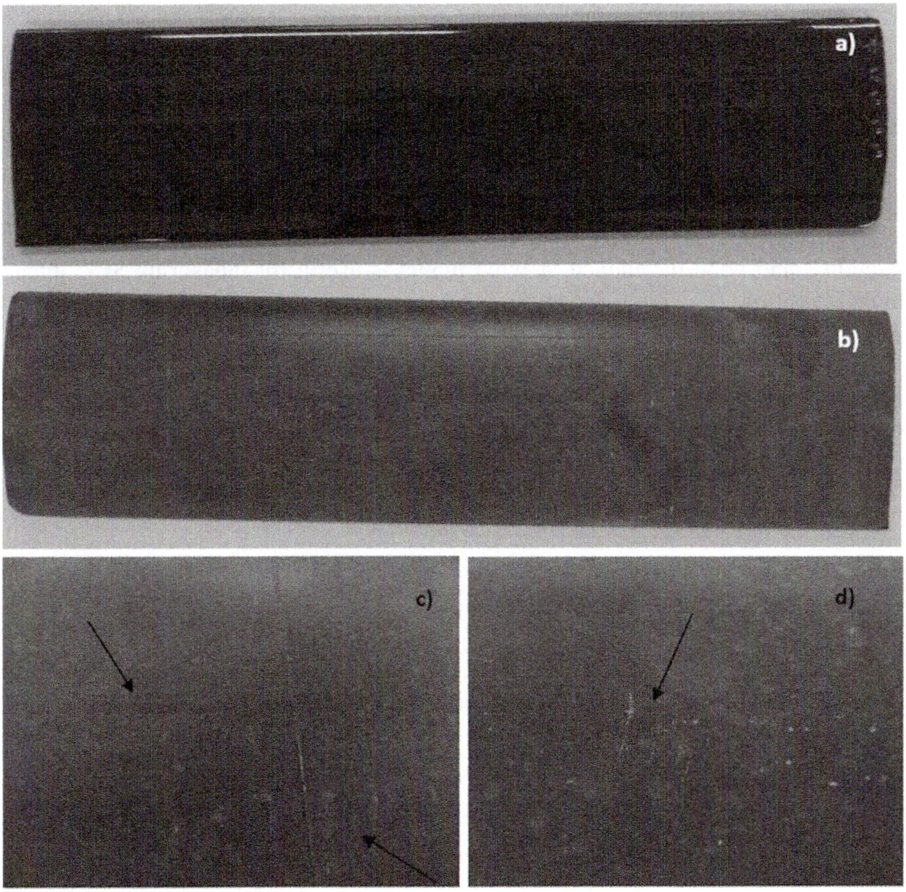

Fig. 9 Aerodynamic profile of an airfoil (**a**) before the test and (**b**) after the test at 30 m/seg, (**c**) showing perpendicular lines and kaolin accumulation, and (**d**) deep marks on the airfoil surface

in Fig. 9c, deeper lines appeared on the extrados profile. This phenomenon is common, because the suspended particles in the wind impact on the attack edge and are deposited near the pressure center, provoking exfoliation and delamination on the upper wing or extrados surface [23]. These effects characteristically predispose the materials to SCC. The aggressive environment of the atmosphere places sustained stress on the primary aircraft structures (airframes, landing gear, wing, and fasteners), exposing them to additional damages such as pitting corrosion, exfoliation, and delamination [24]. Consequently, all airframes present microcracks (inter- and transgranular cracks), macrocracks, chemical waste (originating from humidity and chemical reactions between the material and the environment), and holes [25]. In an operative aircraft, only the primary structure is monitored, so damages are detected by nondestructive test methods, because it is the form to monitoring the primary structure [26]. However, in overhaul procedures, individual parts that are corrosion damaged or have outlived their useful lifetime are replaced [27].

4 Conclusion

The preliminary results confirm that SCC damage can be detected in the designed wind tunnel. SCC is sourced from shock of suspended particles in the atmosphere, humidity, and wind turbulence. A wind tunnel is useful for measuring SCC under flight conditions (in which aircraft are exposed to sustained turbulent flow, particles, and humidity).

Acknowledgments The authors acknowledge Alonso Romero-Jabalquinto and Miguel Angel Neri-Alvarez for their invaluable support during the first stage of the process design.

Future Work In the next stage of our research, we will measure the stress and determine the wind corrosion under flight conditions.

References

1. Roberge, P. R. (2008). *Corrosion engineering, principles and practice* (1st ed., pp. 197–201). New York: McGraw Hill.
2. Doung, C. N., & Wang, C. H. (2007). *Composite repair, theory and design* (1st ed., pp. 12–14). Amsterdam/Boston: Elsevier.
3. Benavides, S. (2009). Corrosion in aerospace industry. In *Corrosion control in the aerospace industry, Woodhead publishing in materials* (pp. 1–13). Boca Raton: CRC Press.
4. Baker, A., Dutton, S.,& Kelly, D. (2004). *Composite for aircraft structures. AIAA education series* (2nd ed., pp. 113–115, 191–203). Woodhead Publishing, Reston, VA, USA.
5. Cantor, B., Assender, H., & Grant, P. (2001). *Aerospace materials* (Vol. 1). Institute of Physics Publishing. Philadelphia, PA, USA.
6. Hou, W., Zhang, W., Liu, X., & Wang, Z. (2011). Ding. M.: Failure analysis of aviation torsional springs. *Chinese Journal of Aeronautics, 24*(4), 527–532.
7. Zhang, Q., Song, F. B., Wang, H. G., & Luo, J. H. (2002). Influence of ambient–temperature corrosion on the oxidation behavior of al-Si coatings. *Chinese Journal of Aeronautics, 15*(4), 244–249.
8. Baker, C. J. (2007). Wind engineering—Past, present and future. *Journal of Wind Engineering and Industrial Aerodynamics, 95*, 843–870.

9. Cermak, J. E. (1979). Application of wind tunnels to investigation of wind-engineering problems. *AIAA Journal, 17*(7), 679–690.
10. Ahmed, N. A. (2013). Design features of a low turbulence return circuit subsonic wind tunnel having interchangeable test sections. In N. A. Ahmed (Ed.), *Wind tunnel designs and their diverse engineering applications* (pp. 29–57). Rijeka: InTech. New South Wales, UK.
11. Hernández, M. A. G., López, A. I. M., Jarzabek, A. A., Perales, J. M. P., Wu, Y., & Xiaoxiao, S. (2013). Design methodology for a quick and low-cost wind tunnel. In N. A. Ahmed (Ed.), *Wind tunnel designs and their diverse engineering applications* (pp. 3–28). Rijeka: InTech. New South Wales, UK.
12. Sastri, S., Ghali, E., & Elboujdaini, M. (2007). *Corrosion prevention and protection, practical solutions* (1st ed., pp. 336–384). Hoboken: Wiley. New York, USA.
13. Anderson, J. (2010). *Fundamentals of aerodynamics, Anderson series* (5th ed., pp. 292–300). McGraw-Hill Education. New York, NY, USA.
14. Barlow, J. B., Rae, W. H., & Pope, A. (1999). *Low-speed wind tunnel testing* (3rd ed., pp. 62–213). New York: Wiley.
15. Mehta, R. D., & Bradshaw, P. (1979). Design rules for small low speed wind tunnel. *The Aeronautical Journal of the Royal Aeronautical Society., 83*, 443–453.
16. Mehta, R. D. (1977). The aerodynamic design of blower tunnels with wide-angle diffusers. *Prog. Aerospace Sci., 18*, 59–120.
17. Mikhail, M. N. (1974). Optimum design of wind tunnel contractions. *AIAA Journal, 17*(5), 471–477.
18. Chmielewski, G. E. (1974). Boundary-layer considerations in the design of aerodynamic contractions. *Journal of Aircraft, 11*(8), 435–438.
19. Cengel, Y., & Cimbala, J. M. (2006). *Fluid mechanics, fundamentals and applications* (2nd ed., pp. 351–357). New York: McGraw-Hill.
20. Bell, J. H., & Mehta, R. (1998). *Contraction design for small low-speed wind tunnels.* Washington, DC: National Aeronautics Space Administration-Ames Research Center.
21. Nordin, N., Ambri, Z., Karim, A., Othman, S., & Raghavan, V. R. (2013). Design and development of low subsonic wind tunnel for turning diffuser application. *Advanced Materials Research, 614–615*, 586–591.
22. Cattafesta, L., Bahr, C., & Mathew, J. (2010). Fundamentals of wind-tunnel design. In R. Blockley & W. Shyy (Eds.), *Encyclopedia of aerospace engineering*. Hoboken: Wiley.
23. Phelps, E. H., & Loginow, A. W. (1960). Stress corrosion of steels for aircraft and missiles. *Corrosion-National of Corrosion Engineers, 16*, 325t–335t.
24. Wanhill, R. J. H., & Byrnes, R. T. (2017). Stress corrosion cracking in aircraft. In *Aerospace materials and material technology, Indian institute of metals series* (pp. 387–410). Singapore: Springer Science + Business, Media.
25. Carter, A. E., & Hanagud, S. (1975). Stress corrosion susceptibility of stress-coined fastener holes in aircraft structures. *AIAA Journal, 13*(7), 858–863.
26. Liao, M., Bellinger, N. C., & Komorowski, J. P. (2003). Modelling the effects of prior exfoliation corrosion on fatigue life of aircraft wing skins. *International Journal of Fatigue, 25*, 1059–1067.
27. Pidaparti, R. M., & Patel, R. R. (2008). Correlation between corrosion pits and stresses in al alloys. *Materials Letters, 62*, 4497–4499.

Citric Acid Passivation of 15-5PH and 17-4PH Stainless Steel Used in the Aeronautical Industry

M. del R. Lara Banda, D.Y. Pérez Ortíz, C. Gaona Tiburcio, P. Zambrano-Robledo, J.A. Cabral Miramontes, and F. Almeraya Calderon

Abstract In the aeronautical industry, the effects of corrosion are reflected in high costs of maintenance and equipment inactivity, as well as in safety risks for personnel. Modern aircraft are manufactured using different types of metal alloys, for example, civil aircraft are constructed with heat-treated aluminum alloys and military aircraft with titanium alloys and stainless steels. The objective of this study is to use electrochemical techniques to determinate the layer passivate of stainless steel (SS) 304, 17-4PH, and 15-5PH. Passivation of the SS was performed in 15% citric acid at temperatures of 25 and 49 °C. The corrosion kinetics was obtained using the electrochemical technique as potentiodynamic polarization curves (PPC) in a three-electrode system: the electrolytes used were sodium chloride (NaCl) and sulfuric acid (H_2SO_4). Passivation in citric acid allows obtain passive layers at temperatures of 49 °C with immersion times of 30 min. Precipitation hardening steels allow to obtain passive layers up to 360 mV in sodium chloride (NaCl), and in sulfuric acid there is a mechanism of passivation-transpassivation-secondary passivation, this due to the high electropositive values of potential above 1000 mV.

Keywords Aeronautics • Stainless steels • Corrosion • Electrochemistry • Passivation

1 Introduction

In the manufacture of helicopters, about 8% of SS is used [1]. The use is limited due to the density that steel exhibits compared to other materials such as aluminum (2.5 denser), titanium (1.5 denser), and composites (3.5 denser). The properties of this

M.d.R. Lara Banda (✉) • D.Y. Pérez Ortíz • C. Gaona Tiburcio • P. Zambrano-Robledo
J.A. Cabral Miramontes • F. Almeraya Calderon
Universidad Autónoma de Nuevo León, UANL. Facultad de Ingeniería Mecánica y Eléctrica, FIME. Centro de Investigación e Innovación en Ingeniería Aeronáutica, CIIIA, Aeropuerto Internacional del Norte, Carretera a Salinas Victoria Km. 2.3, Apodaca, NL 66600, Mexico
e-mail: marialarabanda@yahoo.com.mx

P. Zambrano-Robledo et al. (eds.), *Proceedings of the Symposium of Aeronautical and Aerospace Processes, Materials and Industrial Applications*,
DOI 10.1007/978-3-319-65611-3_9

material have the advantage of wear resistance, corrosion resistance, costs, and availability [1]. Corrosion in commercial, military, and private fuselages is considered a major problem that directly affects safety, economic, and logistical issues [2]. The most common method used to improve corrosion protection in stainless steels is passivation. This term is defined as "treatment or coating generated to reduce the chemical reactivity of its surface" [3, 4].

The passivation layer is an external protection (metal oxide) to reduce the detrimental environmental impact of factors such as air or water. Consequently this process requires an end product that is clean and free from iron and other contaminants [3–6].

The surface of stainless steel is considered passivated, even when exposed to ambient conditions, and while the dissolution of the metal could be expected by thermodynamic considerations, this passive layer remains essentially unchanged over time [7, 8].

Research has been carried out to date to obtain passive agents less toxic to the environment. In 2003 [9], the company Boeing evaluated the use of citric acid as alternative passivation of stainless steel aeronautical parts. Subsequently in 2008 [4] and 2011 [5], the National Aeronautics and Space Administration (NASA) continued with assessment of the use of citric acid as a passivating agent through stress tests, adhesion, and exposure in the salt spray chamber, making a recommendation to continue passive layer evaluation studies. This work aims to study the temperature conditions and passivation times to obtain stable passive layers resistant to corrosion in precipitation hardening stainless steel.

2 Methodology

Materials used in this investigation were SS 304, 15-5PH, and 17-4PH whose chemical composition was determined by X-ray fluorescence.

Preparation of the electrodes: each of the SS was prepared according to ASTM A380 [10] in order to obtain a surface without harshness, after which a washing and rinsing was carried out to eliminate preparation impurities, thus obtaining the material to passivate.

The passivation process was carried under ASTM A967 [11], which consisted of the preparation of a 15% citric acid solution at a constant temperature of 25 and 49 °C (± 2 °C) which the material was introduced to passivate for a time of 30 and 90 min, after which a rinse was performed. The experimentation matrix is shown in Table 1.

Table 1 Experimental matrix of passivation of steels in 15% citric acid

	Temperature (°C)			
	25		49	
	Time (min)			
Stainless steel	30	90	30	90
304	X	X	X	X
15-5PH	X	X	X	X
17-4PH	X	X	X	X

Fig. 1 Corrosion cell three electrodes for PPC

Table 2 Chemical composition of stainless steels (%wt.)

Stainless steel	Fe	Cr	Ni	Mn	Cu	Si
304	Balance	18.06	07.96	01.63	00.36	00.60
15-5PH	Balance	15.04	04.12	00.80	03.25	00.59
17-4PH	Balance	15.03	04.53	00.74	03.49	00.48

Electrochemical system: all experiments were performed by immersion in a corrosion cell, using a three-electrode arrangement, consisting of work electrode (passivated specimen), saturated calomel reference electrode (SCE), and auxiliary electrode of platinum (Fig. 1). The electrochemical technique used was PPC according to ASTM G5 [12] (potential polarization of −1000 to 1200 mV with a velocity polarization of 60 mV/min), with electrochemical technique of current direct enable to observe the corrosion kinetics and consequently the presence of material passivation [13]. A galvanostat/potentiostat/ZRA-Gill equipment from ACM Instruments was used. The electrolytes were NaCl 5% and H_2SO_4 1%.

3 Results and Discussion

The chemical composition obtained by X-ray fluorescence of the SS can be seen in Table 2.

3.1 PPC in Sodium Chloride (NaCl)

Figures 2, 3, and 4 show the PPC obtained in SS 304, 17-4PH, and 15-5PH, passivated in citric acid and exposed in a sodium chloride (NaCl) medium at room temperature (25 °C).

The results were obtained in stainless steel 304 passivated at a temperature of 25 and 49 °C with immersion times of 30 and 90 min (see Fig. 2). In all the cathodic and anodic polarization curves, a mixed control by activation is observed. The anodic curve at 25 °C/90 min presents activation followed by a pseudo-passivation.

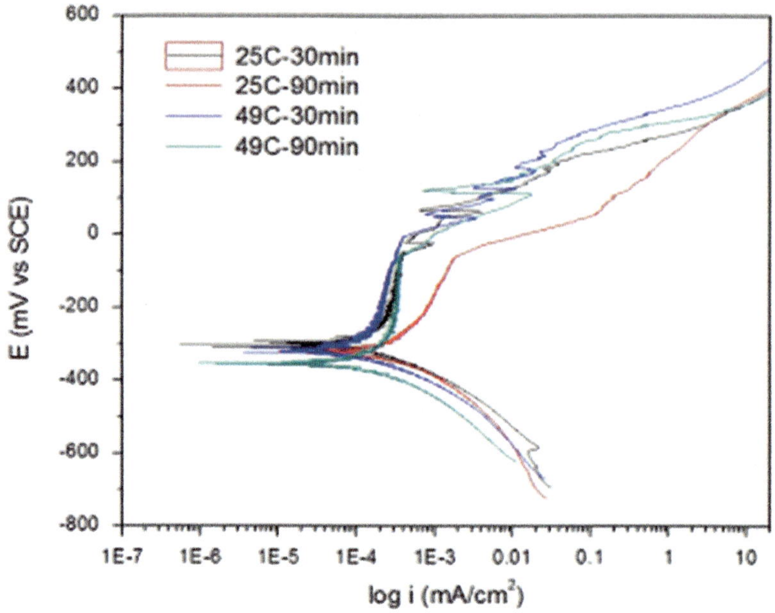

Fig. 2 PPC, for 304 SS passivated in citric acid in NaCl

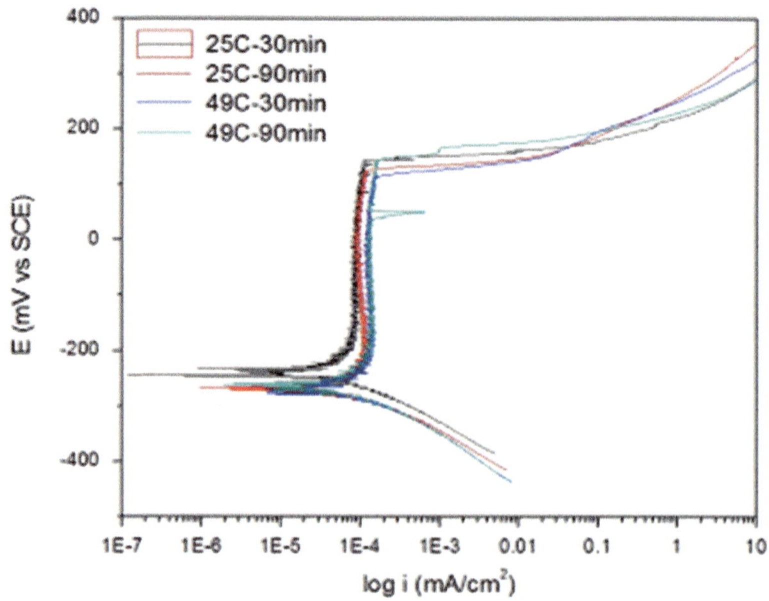

Fig. 3 PPC, for 15-5PH SS passivated in citric acid in NaCl

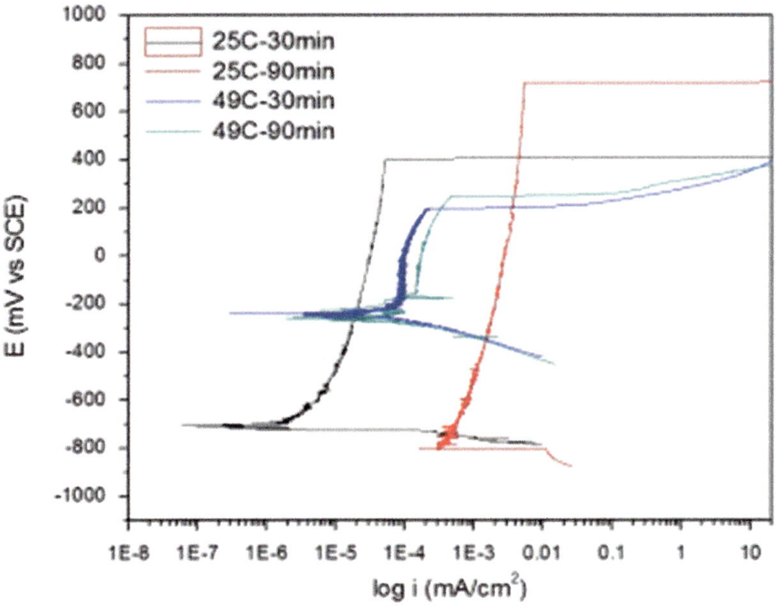

Fig. 4 PPC, for 17-4PH SS passivated in citric acid in NaCl

At this same temperature but with 30 min of immersion in the passivating agent, the anodic curve indicates passive layer formation with breakdown of the same, caused by the dissolution of the metal, presenting a corrosion potential of −298 mV. The temperature curve at 49 °C/30 min shows the same behavior at 25 °C/30 min with a corrosion potential of −318 mV. The passivation temperature obtained at 49 °C/90 min and with a corrosion potential of −352 mV where passive layer formation and dissolution are observed, similar behavior is indicated by the polarization curves of 25 °C/30 min and 49 °C/30 min. Observe that the 25 °C/90 min curve shows an order of magnitude different in the corrosion current density with respect to the curves of 25 °C/30 min and 49 °C/30 min and 90 min which have a passivation zone of 150 mV. The cathode region in all curves shows activation.

Figure 3 shows the PPC for steel 15-5PH at 25 and 49 °C with passivation times of 30 and 90 min; in all the curves, a mixed control is observed in the anodic and cathodic part. At temperatures of 25 °C/90 min and 49 °C/30 and 90 min, there is a displacement of the curves to more active potentials of around −270 mV. The curve of 49 °C/90 min shows a breakdown of the passive layer; this same curve has the best passive layers of 240 mV followed by the curve at 25 °C/90 min with 360 mV. At 25 °C/90 min and 49 °C/30 min, there is transpassivity of the material followed by activation, whereas at 49 °C/90 min and 25 °C/30 min, there is a stepped increase in the corrosion current density.

Anodic control is observed in the PPC of 17-4PH steel at 25 °C/30 min and 90 min (Fig. 4); the cathodic region in both cases is limited by the reduction process.

In both curves there are pitting and regeneration of the passive layer, followed by transpassivity accompanied by dissolution of the material. At 90 min, there is a greater stability of the passive layer of 750 mV. At temperature of 49 °C/30 and 90 min, there is a displacement of the curves toward less active potentials, around −360 to −240 mV. The passive times of 90 min, there is an abrupt change that causes pitting and breakdown of passive layer, caused by the dissolution of the material, and there is passive layer stability of 180 mV. In both cases a process of transpassivation is followed. At this passivation temperature, the curves have a mixed polarization control, where the cathodic zone is observed on activation.

The analysis of PPC in response to a sodium chloride (NaCl) medium shows that austenitic stainless steel 304 at temperature of 49 °C has small passive layer formation of 150 mV. Above −100 mV there is breakdown of layer with system activation. The 15-5PH presents a good response to the polarization generating passive layers of up to 360 mV, followed by a transpassivity. In this material the passivation temperature can be 25 °C but with immersion times of up to 90 min. Steel 17-4PH exhibits passive layer formation of 180 mV at temperatures of 49 °C; at an ambient temperature, the material exhibits passivity but at active corrosion potentials.

3.2 *PPC in Sulfuric Acid (H_2SO_4)*

Figures 5, 6, and 7 show the PPC obtained in SS 304, 15-5PH, and 17-4PH passivated in citric acid and exposed in a sulfuric acid (H_2SO_4) at room temperature (25 °C).

Figure 5 shows the results obtained for steel 304 with a mixed polarization control with a cathodic region in activation with reduction of hydrogen ions. The anodic curve exhibits a limited corrosion densities of 10^{-3} mA/cm^2 at 10^{-2} mA/cm^2 and later exhibits a transition of the active solution, and the passivating film begins to form and causes an abrupt fall of the i_{corr}. The increase of potential to electropositive values generates a pseudo-passive layer up to 375 mV, a potential from which the decomposition of the film takes place generating the oxidation of the material and transpassive region of the material. The disolution is above 1000 mV beginning of a region of secondary passivity, at 49 °C/90 min, and a secondary pseudo-passivation at 49 °C/30 min and 25 °C/30 and 90 min.

In 15-5PH (Fig. 6), a mixed control with corrosion potentials of around −320 mV is observed in polarization curves for the different temperatures and passivation times. With the exception of curve 49 °C/90 min, the others have an anodic behavior with critical passivation current densities of order 10^{-2} mA/cm^2 with a current density drop at potentials of −250 mV, potential in which there is an active solution of the material and initiation of the formation of the protective film, generating decrease of the current density followed by a passive state of −125 to 440 mV. Above this latter potential, decomposition of the film occurs with an increase in the anodic current density generating oxidation of the material and producing insoluble ions and giving rise to the transpassivity. A secondary pseudo-passivation is present

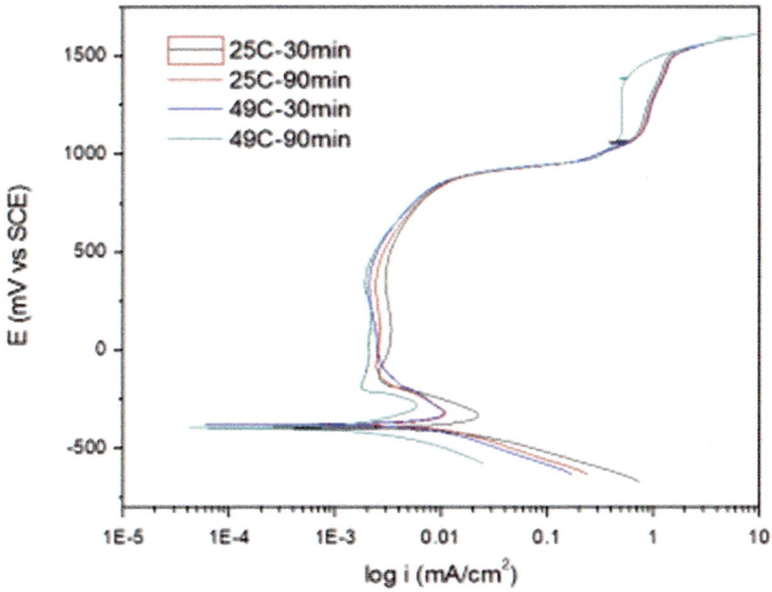

Fig. 5 PPC, for 304 SS passivated in citric acid in H_2SO_4

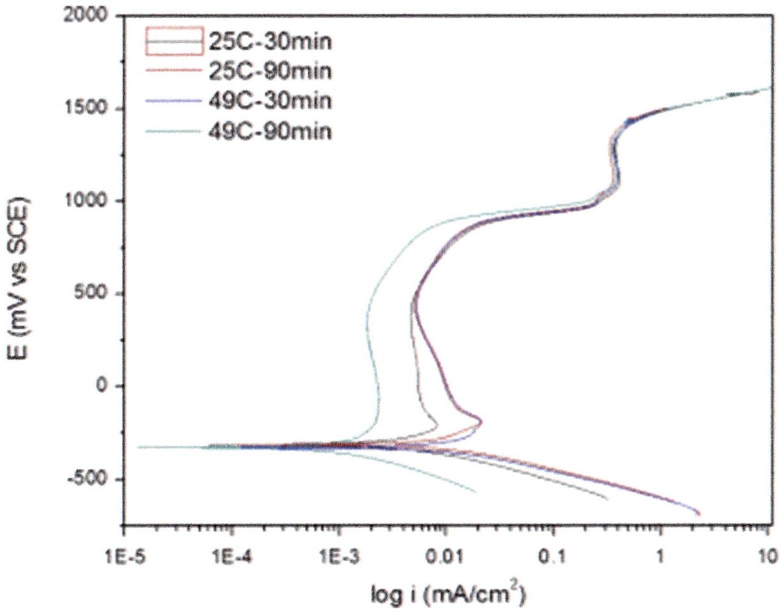

Fig. 6 PPC, for 15-5PH SS passivated in citric acid in H_2SO_4

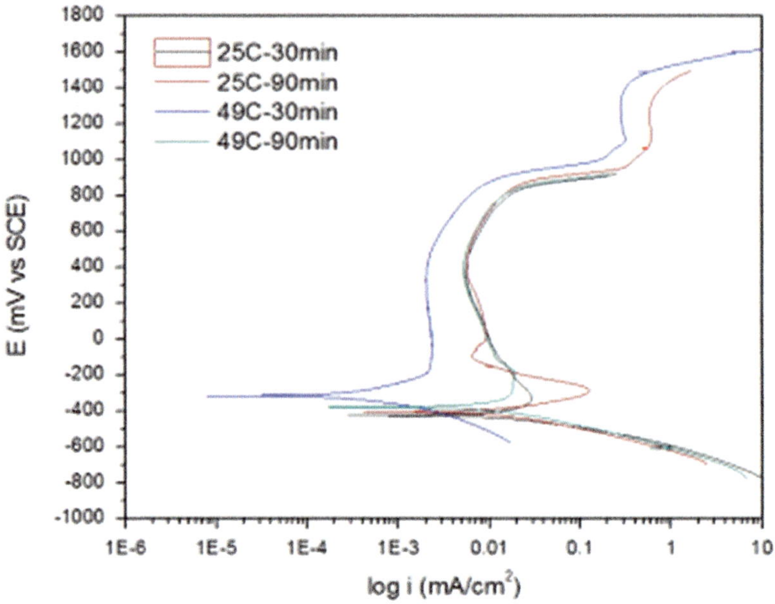

Fig. 7 PPC, for 17-4PH SS passivated in citric acid in H_2SO_4

above a potential of +900 mV until secondary passivation occurs in potentials around 1200 mV and an i_{corr} of the order of 10^{-1} mA/cm^2.

In the PPC (Fig. 7), a mixed control for 17-4PH stainless steel is observed. The cathode region is activated, and the anode region exhibits displacement to electropositive potentials as the passivation temperature increases from 25 to 49 °C. The corrosion potential for the 49 °C/30 min curve is −318 mV with a passive layer of 600 mV followed by transpassive interval from 420 to 1000 mV, potential in which the degradation of the film is observed; increasing the i_{corr} to values potential above 1000 mV, a secondary pseudo-passivation is observed; and when it achieves values of 1100 mV, there is presence of secondary passivation. A similar behavior occurs in the curve at 49 °C/60 min and 25 °C/30 min. The curve 25 °C/90 min presents an abrupt drop of the i_{corr} after the beginning of the formation of the passive layer.

PPC performed on SS passivated in citric acid at different temperatures and times and exposed in sulfuric acid (H_2SO_4) present a critical current density at potentials of −330 mV, with the beginning of passive layer formation, followed by an active solution that generates transpassivity and formation of a secondary passivation. Only precipitated hardening stainless steel at 49 °C shows passive layer formation with lower current demand.

The electrochemical parameters obtained from the PPC are presented in Table 3. The corrosion rates 10^{-7} mm/year are observed for sodium chloride (NaCl) solution, while for sulfuric acid corrosion rates 10^{-4} mm/year, having three orders of magnitude more resistant in sodium chloride, within which 17-4PH present values of 10^{-8} mm/year.

Table 3 Electrochemical parameters obtained from the PPC

Stainless steel	Electrolyte	Temperature (°C)	Time (s)	Ecorr (mV)	Rp (Ω/cm²)	Icorr (mA/cm²)	Corrosion rate (mm/year)
304	Sodium chloride 5%	25	30	−298.52	251808.94	8.51E-05	8.70E-07
			90	−323.26	135432.89	1.39E-04	1.42E-06
		49	30	−318.96	327808.84	6.24E-05	6.38E-07
			90	−351.97	261644.37	6.41E-05	6.56E-07
	Sulfuric acid 1%	25	30	−395.35	2599.29	1.04E-02	1.06E-04
			90	−379.14	4629.07	5.17E-03	5.29E-05
		49	30	−380.08	5911.22	5.52E-03	5.65E-05
			90	−394.42	18440.89	1.36E-03	1.39E-05
15-5PH	Sodium chloride 5%	25	30	−245.82	470472.76	3.32E-05	3.50E-07
			90	−268.49	351739.03	4.23E-05	4.46E-07
		49	30	−270.37	335147.65	5.00E-05	5.26E-07
			90	−266.34	359877.03	4.50E-05	4.75E-07
	Sulfuric acid 1%	25	30	−312.68	9891.06	3.24E-03	3.41E-05
			90	−312.71	4327.23	6.09E-03	6.41E-05
		49	30	−328.68	3607.43	7.73E-03	8.15E-05
			90	−328.25	37421.13	6.55E-04	6.90E-06
17-4PH	Sodium chloride 5%	25	30	−708.51	2016194.75	4.48E-06	4.68E-08
			90	−349.21	204859.00	8.73E-05	9.12E-07
		49	30	−239.22	473828.14	3.23E-05	3.38E-07
			90	−360.17	343896.74	5.38E-05	5.63E-07
	Sulfuric acid 1%	25	30	−422.23	2004.83	1.11E-02	1.16E-04
			90	−405.16	1822.75	1.08E-02	1.12E-04
		49	30	−318.50	76227.32	3.17E-04	3.31E-06
			90	−376.42	3193.16	8.61E-03	9.00E-05

4 Conclusions

- 15-5PH SS has passive layers stable up to 360 mV. The passivation temperature is a determining factor in obtaining the passive layer at 49 °C, where a better layer with a 30 min immersion time is obtained in the passivating agent. In the chloride environment, passivation is observed followed by a zone of transpassivity, while in sulfuric acid, there is passivation-transpassivation-secondary passivation. In sulfuric acid there are electropositive potentials up to 1000 mV, indicating passive layer regeneration.
- Results indicate that 17-4PH SS achieves corrosion rate of 10^{-8} mm/year in sodium chloride at temperatures of 25 °C/30 min, a condition in which there is a stable passive layer of 800 mV.
- The 304 SS at the different passivation conditions presents pseudo-passivation and passive layer breakdown. For the above, unstable passive layers are achieves.
- In precipitation hardening stainless steel, passivation in citric acid allows to obtain passive resistant layers in the presence of sodium chloride (NaCl) and sulfuric acid H_2SO_4.

References

1. Mouritz, A. P. (2012). *Introduction to aerospace materials* (p. 233). Oxford: Woodhead Publishing.
2. Siddiqui, T. (2015). *Aircraft materials and analysis* (p. 127). New York: Mc Graw Hill.
3. Lopes, J. (2008). Material selection for aeronautical structural application. *Ciência & Tecnologia dos Materiais, 20*, 78–82.
4. Lewis, P. L., Kolody, M., & Carran, J. (2008). *Alternative to nitric acid passivation of stainless steel alloys*. Presented at NASA Technology Evaluation for Environmental Risk Mitigation Principal Center (TEERM). NASA John F. Kennedy Space Center, FL.
5. Yasenski, D., Reali, C., Larson, J. (2011). *Citric acid passivation of stainless steel*. Presented at the aircraft Airworthiness and Sustainment Conference. USA.
6. Olsson, C., & Landolt, D. (2003). Passive films on stainless steels-chemistry. *Structure and Growth. Electrochimica Acta, 48*, 1093–1104. doi:10.1016/S0013-4686(02)00841-1
7. Gaydos, S. P. (2003). *Passivation of aerospace stainless steel parts with citric acid solutions*. St Louis Missouri. USA: The Boeing Company.
8. Gordon, J. (1986). *An introduction to stainless steel* (pp. 1–5). American Society for Metals International (ASM) SM, USA.
9. Cramer S.D., Covino B.S. (1992). *Handbook. Corrosión: Materials* (Vol. 13B, pp. 54–77). American Society for Metals International (ASM), USA. doi:10.1361/asmhba0003812.
10. ASTM A-380. (1999). *Standard practice for cleaning, descaling and passivation of stainless steel parts, equipment, and systems*. ASTM Standard, USA.
11. ASTM A967. (1999). *Standard specification for chemical passivation treatments for stainless steel parts*. ASTM Standard, USA.
12. ASTM G5. (1999). Standard reference test method for making potentiostatic and potentiodynamic anodic polarization measurements. ASTM Standard, USA.
13. Genescá, J., & Ávila J. (1996). *Más allá de la Herrumbre III-Corrosión y Medio Ambiente* (pp. 45–46). México, D.F: Fondo de Cultura económica.

Part III
Numerical Modelling

Grain Size Effect during Hot Ring Rolling Process Using Nickel-Based Superalloy

J.M. Cisneros, L.A. Reyes-Osorio, P. Zambrano-Robledo, and F. Montemayor-Ibarra

Abstract The components of aircraft engines require high strength materials at high temperatures and corrosion. The nickel-based superalloys possess mechanical properties that are directly dependent on grain size; the aeronautical components with coarse grain size are usually employed to minimize creep deformation, while those with finer grain size have the mission to increase the fatigue resistance. One of the advantages of the ring rolling process is the favorable grain flow and good surface quality of the rolled rings. This paper shows the numerical simulation of the hot ring rolling process, where two initial grain sizes are used, one of 20 μm, this being the fine grain size, and 80 μm for analysis of coarse grain size. Both simulations are conducted at an initial temperature of 1100 ° C. At the conclusion of the process, both simulations presented a refinement in grain size, proving the efficiency of rolling.

Keywords Ring rolling • Inconel 718 • Grain size

1 Introduction

The ring rolling process has been widely used in the production of seamless rolled rings, which are commonly used in the aerospace industry [1]. The process consists of three movements in its tools, starting with the mandrel that presses the walls of the workpiece against the main roller. This main roller has rotation speed, whereas the mandrel and the axial rollers only have torque movement [2, 3]. Figure 1 shows the tools used in the process with their respective movements.

J.M. Cisneros (✉) • L.A. Reyes-Osorio • P. Zambrano-Robledo • F. Montemayor-Ibarra
Universidad Autónoma de Nuevo León. Facultad de Ingeniería Mecánica y Eléctrica. Centro de Investigación e Innovación en Ingeniería Aeronáutica, Ave. Universidad S/N, Cd. Universitaria, San Nicolás de los Garza, NL C.P. 66451, Mexico
e-mail: miguelcisnerosv@gmail.com

P. Zambrano-Robledo et al. (eds.), *Proceedings of the Symposium of Aeronautical and Aerospace Processes, Materials and Industrial Applications*,
DOI 10.1007/978-3-319-65611-3_10

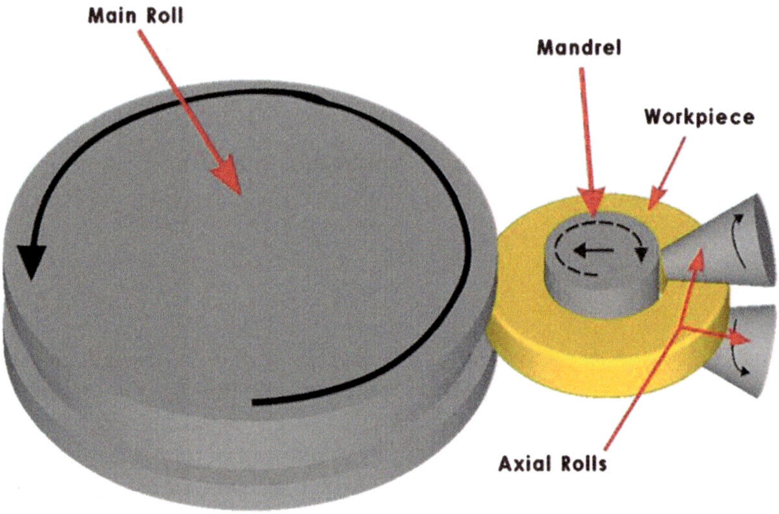

Fig. 1 Schematic of the ring rolling process

Table 1 Chemical composition of Inconel 718 (wt. %) [4]

Ni	*Cr*	*Fe*	*Nb*	*Mo*	*Ti*	*Co*	*Al*	*Si*	*Mn*
55	18	17.3	5.4	2.95	0.91	0.24	0.68	0.08	0.07
B	*C*	*Mg*	*P*	*S*	*Pb*	*Ca*	*Se*	*Bi*	*Cu*
0.006	0.08	0.0033	0.007	0.001	0.0001	<0.1	<0.0001	<0.0001	0.07

The Inconel alloys are one of the most used materials for the ring rolling process in aerospace; these alloys have a nickel and chrome base that act together to resist oxidation, carburization, and other forms of deterioration at high temperatures. The Inconel superalloys do not become brittle at cryogenic temperatures, possess good tensile strength and fatigue at moderate temperatures, and exhibit excellent creep resistance at high temperatures; therefore, their applications have a long reach, since they are used in fields of aviation and aerospace and also mainly in nuclear power and petrochemical industries [4]. Since the microstructure of this alloy after rolling is heavily affected by the rolling speed and badly influences the properties of the product, study on the microstructure evolution during industrial hot rolling is important for steel industry to predict the mechanical properties of the hot-rolled products and design the most effective processing path to fulfill the quality requirements [5, 6]. Table 1 shows the chemical composition of the alloy used in the simulation.

The finite element method is a numerical technique, which emerged as such in the 1960s, and since then to date has gone from being a novel tool to being necessary both academically and industrially. They have been applied in almost all fields of knowledge with success, reaching their major applications in the field of engineering. The success of the method is also due to the emergence of faster computers, with greater storage capacity and data management, coupled with the development of specialized programs. In recent years, many of the advances in the ring rolling process were obtained mainly by numerical simulations performed with FEM program packages. For example, Huez et al. [7] developed a three-dimensional model and observed the microstructural deformations of the ring, Wang et al. [8] developed a virtual radial-axial ring lamination process to guide the design and optimization of processes using the LS-DYNA FE code, and Yeom et al. [9] investigated the radial-axial design of a large-scale ring product of Ti-6Al-4 V alloy using an analysis by FEM. From the forming principle of radial-axial ring rolling (RARR) above, it can be seen that RARR is an extremely complex dynamic rolling process with high flexibility. Recently, researches on RARR are mainly focused on FE modeling method and process simulation. For instance, Xie et al. [10] developed a rigid-viscoplastic dynamic explicit FE code to simulate ring rolling. Davey and Ward (2003) [11]proposed a practical method for FEM simulation of ring rolling using ALE flow formulation instead of conventional Lagrangian approach. Kim et al. [12] put forward a dua-mesh approach to ring rolling simulation with high computational efficiency. By FEM simulation, deformation behaviors during ring rolling process and effects of process conditions on ring rolling have been revealed systematically. Moreover, FEM simulation of microstructure evolution during ring rolling has also been explored, especially for some special metals, as superalloys. Besides, macro-microscopic deformation laws during blank forging and hot ring rolling process have been investigated [6, 7]. This paper investigates the grain size change that occurs during ring rolling process, for which two simulations were developed applying the mathematical model of Avrami, one with a fine grain size and the other with an initial coarse grain size, in order to observe recrystallization in both cases.

2 Numerical Model

Two numerical models were developed in the DEFORM 3D finite element program considering the parameters shown in Table 2. The variable studied in both models was the initial grain size; in the first simulation, a fine grain size was considered and

Table 2 Parameters adopted in the numerical simulation

Parameter	Value
Recrystallization volume fraction	0.5
Fine grain | initial grain size (μm)	20
Coarse grain | initial grain size (μm)	80
Friction coefficient μ	0.7
Half of the cone angle of the axial rolls θ (°)	15
Initial temperature of the ring (°C)	1100
Rotational speed of the main roll (rad/s)	1
Translational speed of the mandrel (mm/s)	0.35

in the second one, a coarse grain size. The dimensions of the workpiece used by the author Huez et al. [7] were taken for reference, which are shown in Fig. 2.

In the structured mesh, a global refinement was used in the workpiece, to observe microstructural changes throughout the entire ring. The tools (main roll, mandrel, and axial rolls) were considered as non-deformable rigid bodies; the dimensions of these tools are shown in Table 3. Figure 3 shows the 3D meshing used in the simulation model and its characteristics are described in Table 4.

The ring diameter increases when the pressure applied reduces the wall thickness, and the process continues until the required dimensions are obtained. The height of the ring is controlled by the shape of the main roll and by means of the axial rollers located along the ring. The translation velocity of the mandrel was 0.35 mm/s, and the rotational speed of the main roll was 1 rad/s. Due to the non-lubrication of the ring rolling process tools, the coefficient of friction is high;

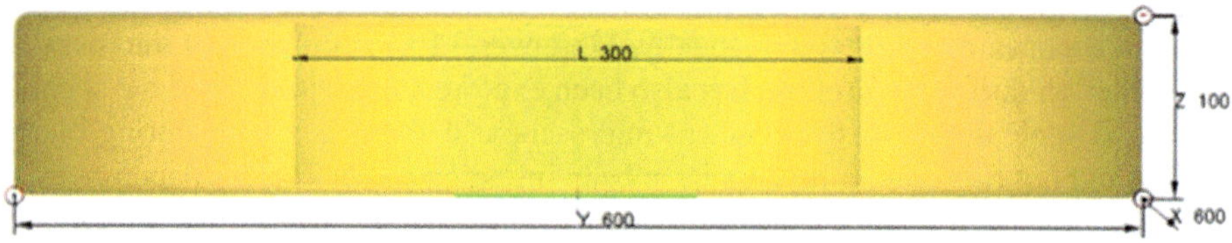

Fig. 2 Initial ring size Dimension of the initial ring workpiece

Table 3 Dimensions of the ring rolling tools

Parameter	Value
Height of the main roll (mm)	240
Radius of the main roll (mm)	600
Radius of the mandrel (mm)	140
Height of the mandrel (mm)	240

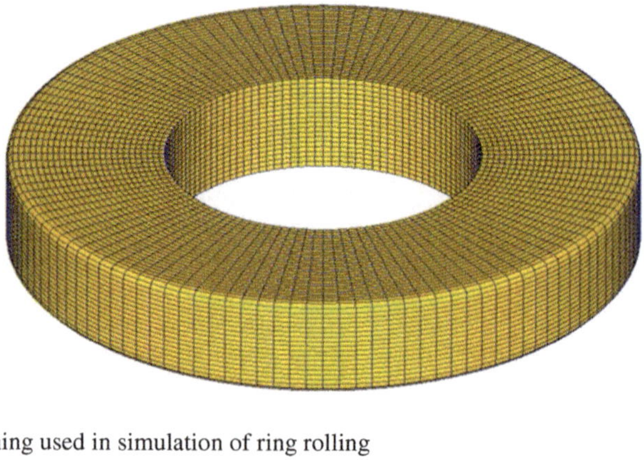

Fig. 3 3D meshing used in simulation of ring rolling

Table 4 3D meshing characteristics

Mesh type	Tetrahedral
Number of elements	59,040
Number of nodes	64,560
Surface polygons	10,800
Revolution sections	120
Remeshing method	Global

Fig. 4 Initial state of hot ring rolling process

therefore, "shear 0.7" was assigned [13]. The recommended temperature range for working the 718 alloy is 900–1200 ° C [14, 15]. For this reason, the initial temperature of the ring was at 1100 ° C, in order to be near the delta phase, which is necessary to produce grain refinement. The complete model of ring rolling process is shown in the Fig. 4.

3 Microstructural Model

The Avrami model [16, 17] was used to evaluate the microstructural change; the general form of the Avrami equation for recrystallized fraction as a function of time is:

$$X = 1 - \exp\left(-kt^{n}\right) \tag{1}$$

In this equation, k and n constants are dependent on material, temperature, previous deformation, and initial grain size [7, 18].

To evaluate the grain size after the deformation, we used Eq. 2, where t is the time retention and n is the material constant, also known as the Avrami exponent, which depends on the nucleation rate and the rate of increase [16, 19]. The parameters used in Eq. 2 are described in Table 5.

Table 5 Parameters of grain growth equation

Parameter	Value
Q_9	390,753
a_9	1.58e + 16
m	2

$$d = \left(d_0{}^m + a_9 t exp\left(-\frac{Q_9}{RT} \right) \right)^{1/m} \quad (2)$$

4 Discussion

The process time was 18 s, 690 mm of outer diameter were assigned as limiting in the simulation, and the walls of the ring were uniformly reduced as were the height. There were no cracks or abnormalities in the final geometry of the ring. Figure 5 shows the dimensions obtained at the end of the process.

Four reference points were taken into account to follow during the simulation; the first point was located at the top of the ring, where the workpiece is in contact with the axial roller. The point number 2 was placed on the inner wall of the ring; in this area the workpiece is in contact with the mandrel, which is the tooling that exerts more force on the ring. In order to follow the inner zone of the ring, a third point was placed; in this area there is no contact with any tool; therefore, it is considered a neutral or control point. The fourth and last point was placed in the contact area of the main roll. In Fig. 6 the points described above are appreciated.

Figure 7 shows the ring at the end of the process, where the small range between the different sizes of grain is highlighted. The average grain size is in the range of 9.53–10 μm. From the results obtained, it can be concluded that the average grain size at the end of the ring rolling is completely uniform, since the value of 10 μm is present in almost 100% of the ring; therefore, the recrystallization occurred homogeneously. The graph of the evolution of the grain size refinement with the four reference points placed in the ring and the area that takes longer to refine is that which is in contact with the main roll, while the region located in the zone of contact with the mandrel achieves a faster recrystallization. The time differences of recrystallization in the different zones of the ring are very small, and they manage to stabilize and remain constant from the 8 s of the process.

In the simulation where the coarse grain size was assigned, values of 39.9–40 μm were obtained; therefore, in the same way that in the analysis of fine grain size, it can be concluded that uniform and homogeneous recrystallization occurs.

For comparing the fine and coarse grain size, the central point was taken as reference, in order to see the evolution of the recrystallization at different grain sizes, Fig. 8 shows the comparison between both grain size. In the cases presented, the recrystallization occurs quickly, since in both simulations, the recrystallization

Fig. 5 Finale ring size of the ring rolling simulation

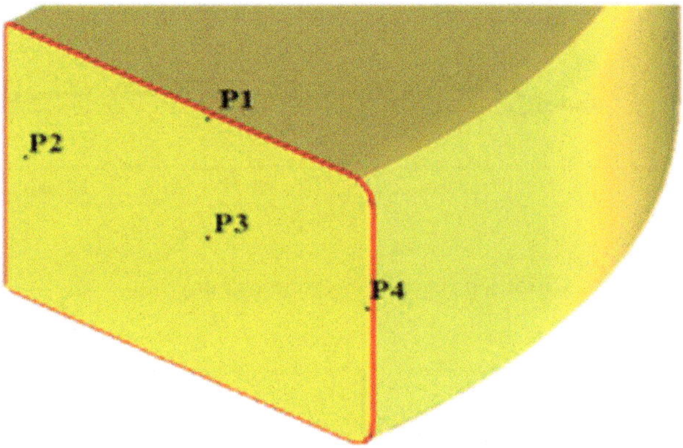

Fig. 6 Relevant points of interest in the simulation

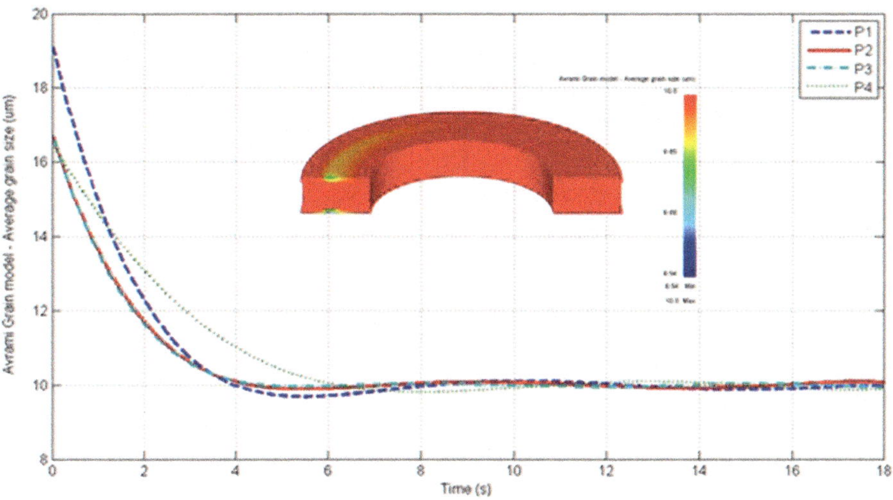

Fig. 7 Evolution of grain size during ring rolling (fine-grain)

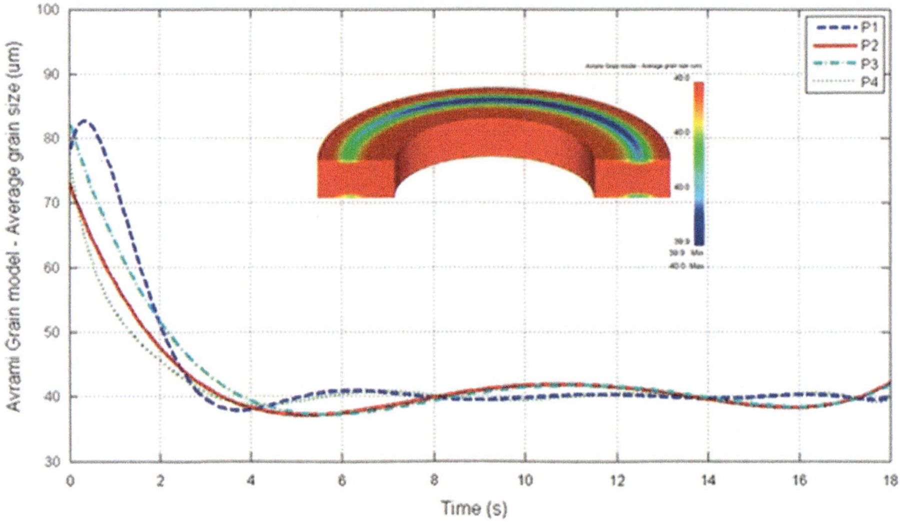

Fig. 8 Evolution of grain size during ring rolling (coarse-grain)

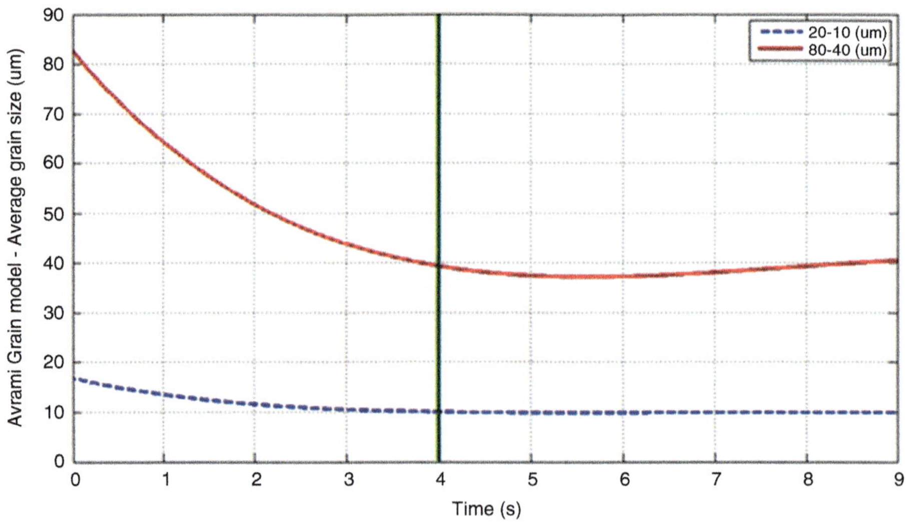

Fig. 9 Comparative graph of grain refinement during ring rolling process

happens in the first seconds of the process. The graph of Fig. 9 shows how fast refinement occurs in the first few seconds, which stabilizes and remains constant until the end of the process.

At the end of the ring rolling, in the experimental work performed by Huez et al. [7], a uniform refinement was obtained throughout all the ring; the main variations of the grain size were presented in the contact zone of the axial rollers, as

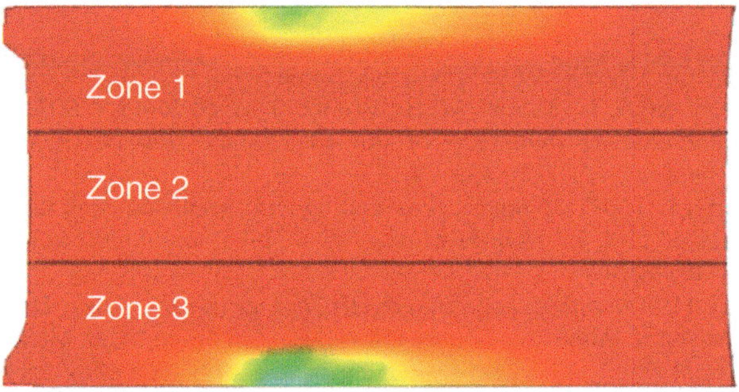

Fig. 10 Reference zones for the comparison of grain sizes

Table 6 Comparison of grain sizes obtained in experimentation and simulation of ring rolling process

Ring zone	Experimental (μm)	Simulation fine grain size (μm)	Simulation coarse grain size (μm)
Zone 1	11–20	9.53–10	39.9–40
Zone 2	11	10	40
Zone 3	11–20	9.53–10	39.9–40

in the simulation performed in this paper. In central and inner regions of the ring a uniform and homogeneous refinement were obtained, The experimentation and the simulation of ring rolling achieved a similar behaviour. In order to compare grain sizes obtained both in the simulations and in the experiment performed by Huez et al. (2001), three zones were assigned to the ring; these zones are shown in Fig. 10. Table 6 shows a comparison of the grain sizes present at the end of the ring rolling process in the experiments carried out by Huez et al. (2001) and the two simulations developed in this work.

5 Conclusions

It was possible to create a finite element model for thermomechanical and microstructural analysis of superalloy 718 in the ring rolling process. The mathematical model of Avrami can predict the microstructural changes during the rolling process; this model allows an estimate and a more precise control of the changes of microstructures that take place during the formation of the metal. Regardless of the initial grain size, recrystallization will occur at the end of the process.

Acknowledgments J. M. Cisneros is grateful for the financial support granted by PAICYT program. The authors wish to thank the Centro de Investigación e Innovación en Ingeniería Aeronáutica of the Facultad Ingeniería Mecánica y Eléctrica of the Universidad Autónoma de Nuevo León.

References

1. Song, J. L., Dowson, A. L., Jacobs, M. H., Brooks, J., & Beden, I. (2002). Coupled thermo-mechanical finite-element modelling of hot ring rolling process. *Journal of Materials Processing Technology, 121*(2-3), 332–340.
2. Guo, L., & Yang, H. (2011). Numerical modelling and simulation of radial-axial ring rolling process. in Chapter 17th. In *Numerical analysis – Theory and application*. Rijeka: In Tech. ISBN 978-953-307-389-7.
3. Ruhrpumpen, T. (1997). *Introduction into the technology of ring rolling*. Dortmund: Thyssen.
4. Loyda, A., Hernández-Muñoz, G., Reyes-Osorio, L., Zambrano, P., & Montemayor-Ibarra, F. (2016). Numerical simulation of rotary forging inconel 718 superalloy applied to aeronautical components. *MRS Proceedings, Symposium 6B, "Advanced Structural Materials", 1812*, 101–107. doi:10.1557/opl.2016.25.
5. Sui, F. L., Zuo, Y., Liu, X. H., & Chen, L. Q. (2013). Microstructure analysis on IN 718 alloy round rod by FEM in the hot continuous rolling process. *Applied Mathematical Modelling, 37*(20), 8776–8784. doi:10.1016/j.apm.2013.04.005.
6. Höllig, K. (2003). Finite element methods with B-splines. *Society for Industrial and Applied Mathematics*. doi:10.1137/1.9780898717532.
7. Huez, J., Noyes, J. L., & Coupu, J. (2001). Three-dimensional finite-element simulation of hot ring rolling. In *Superalloys 718, 625, 706 and various derivatives: Proceedings of the international symposium on superalloys 718, 625, 706 and various* (pp. 625–706). Warrendale: TMS.
8. Wang, Z. W., Zeng, S. Q., Yang, X. H., & Cheng, C. (2007). The key technology and realization of virtual ring rolling. *Journal of Materials Processing Technology, 182*(1), 374–381. doi:10.1016/j.jmatprotec.2006.08.020.
9. Yeom, J. T., Kim, J. H., Park, N. K., Choi, S. S., & Lee, C. S. (2007). Ring-rolling design for a large-scale ring product of Ti–6Al–4V alloy. *Journal of Materials Processing Technology, 187*, 747–751. doi:10.1016/j.jmatprotec.2006.11.042.
10. Xie, C., Dong, X., Li, S., & Huang, S. (2000). Rigid–viscoplastic dynamic explicit FEA of the ring rolling process. *International Journal of Machine Tools and Manufacture, 40*(1), 81–93. doi:10.1016/S0890-6955(99)00043-7.
11. Davey, K., & Ward, M. J. (2003). An ALE approach for finite element ring-rolling simulation of profiled rings. *Journal of Materials Processing Technology, 139*(1), 559–566. doi:10.1016/S0924-0136(03)00537-5.
12. Kim, B., Moon, H., Kim, E., Choi, M., & Joun, M. (2013). A dual-mesh approach to ring-rolling simulations with emphasis on remeshing. *Journal of Manufacturing Processes, 15*(4), 635–643. doi:10.1016/j.jmapro.2013.05.002.
13. Loyda, A., Hernández-Muñoz, G. M., Reyes, L. A., & Zambrano-Robledo, P. (2016). Microstructure modeling of a NiFe based Superalloy during the rotary forging process. *Journal of Materials Engineering and Performance*. doi:10.1007/s11665-016-2104-6.
14. Campbell, F. (2006). *Manufacturing Technology for Aerospace Structural Materials*. Amsterdam: Elsevier.
15. Srinivasan, N., & Prasad, Y. V. R. K. (1994). Microstructural control in hot working of IN-718 superalloy using processing map. *Metallurgical and Materials Transactions A., 25*(10), 2275–2284. doi:10.1007/bf02652327.
16. Cardenas, J., Guajardo, H., Harwood, C., & Manriquez, J. A. (2005). Effects of large reductions and heating temperature-times on grain size control of alloy 718 rolled rings. In *Proceedings of the international symposium on superalloys 718, 625, 706 and various derivatives* (pp. 119–133). Warrendale: TMS.
17. Thomas, J. P., Bauchet, E., Dumont, C., & Montheillet, F. (2004). EBSD investigation and modeling of the microstructural evolutions of superalloy 718 during hot deformation. In K. A. Green (Ed.), *Proceedings of superalloys 2004* (pp. 959–968). Warrendale: TMS.

18. Huang, D., Wu, W. T., Lambert, D., & Semiatin, S. L. (2001). *Computer simulation of microstructure evolution during hot forging of Waspaloy and nickel alloy 718, Proceedings of microstructure modeling and prediction during thermomechanical processing* (pp. 137–147). Warrendale: TMS.
19. Horsinka, J., Kliber, J., Knapinski, M., & Mamuzic, I. (2012). *Study of the kinetics of static recrystallization using avrami equation and stress relaxation method.* In: Conference proceedings WOS, Metal 2012, 277-281. ISBN 978- 80-87294-31-4.

Finite Element Modeling of Friction Stir Spot Welding (FSSW) of Ti-6Al-4V

F. Quiroz, F.A. García-Castillo, L.A. Reyes-Osorio, and A. Ávila-Cabrera

Abstract Titanium alloys, especially Ti-6Al-4V, have been widely used in the aerospace industry due to their high specific strength, high heat-resistant, and high corrosion-resistant properties. This alloy is used in current airframe applications for medium-sized forgings in high load transfer components such as the engine pylon and undercarriage support fittings on the wing, adopting the friction stir spot welding to reduce the use of bolts on structural components. Based on finite element analysis, the three-dimensional model, described in this work, combines the mechanical action of the shoulder and the thermomechanical effect of the welded. Johnson–Cook model is used as a basis of constitutive law in which the yield strength is affected by the temperature and strain rate for the Ti-6Al-4V alloy. The sheet of 2.3 mm thickness were friction stir spot welded using a tool with a convex scrolled shoulder made of a polycrystalline cubic boron nitride (PCBN). The parameter processes employed in this study are rotational speed, plunge speed, and dwell time and plunge depths. The material flow was also numerically simulated using the DEFORM-3D FE software. Results indicated with the incremental penetration of the pin; material from below the pin was displaced upward along the sides of the pin. FE results were verified by experimental results.

Keywords Ti-6Al-4 V • FSSW • FEM

1 Introduction

The FSSW process is a recently developed process and has been studied for applications in the automotive industry and aeronautics among others [1]. The FSSW process is a process derived from the FSW process widely used for joints of aluminum alloys; the FSSW process has the essential difference that there is no transverse movement of the tool [2, 3].

F. Quiroz (✉) • F.A. García-Castillo • L.A. Reyes-Osorio • A. Ávila-Cabrera
Universidad Autónoma de Nuevo León, Facultad de Ingeniería Mecánica Eléctrica. Centro de Investigación e Innovación en Ingeniería Aeronáutica (CIIIA), Ave. Universidad S/N, Cd. Universitaria, San Nicolás de los Garza, NL C.P. 66451, Mexico
e-mail: quiroz.fernando95@gmail.com

P. Zambrano-Robledo et al. (eds.), *Proceedings of the Symposium of Aeronautical and Aerospace Processes, Materials and Industrial Applications*,
DOI 10.1007/978-3-319-65611-3_11

The friction stir welding process is a solid-state process [4, 5], applied with great success in Al, Mg, Cu, Fe, and Ti alloys [6]. Friction stir spot welding (FSSW) is a variation of friction stir spot welding (FSW) welding. The FSSW process consists of three distinct stages: (a) penetration, (b) agitation, and (c) regression. Fig. 1 describes the process which starts with the tool by slowly penetrating the material, which means that the combined effects of the rotation of the tool and the movement of the same penetrating the material promote agitation, and finally the tool is retracted. The process has three main variables, which are the speed of rotation given in RPM, the penetration speed (mm/mm), and the holding time in seconds [7, 8].

The tool penetrates the sheets generating heat by the friction that occurs as the contact increase; when reachs a certain depth, the tool is kept rotating a holding time. The tool shoulder and pin lead to agitation and bonding of the material, and then the tool is retracted giving the end to the welding point [9].

Titanium is a material that has aroused greater interest in the aeronautical sector. The bonding of titanium components by the melt-brazing process frequently degrades the mechanical properties in relation to the base material [10].

The aeronautics industry continually seeks to improve aircraft, due to the high cost of fuel and crew safety [11]; the friction stir spot welding process is a recently developed process that has been studied primarily for applications in this industry; combined with the numerical simulation methods is a fast and efficient way to predict the metallurgical process change. Figure 2 shows the Bridgeport high-speed machining center and the Fluke® Ti400 thermal imager, which were used to weld the joins and obtain the temperature profile, respectively.

Fig. 1 FSSW friction stir spot welding process

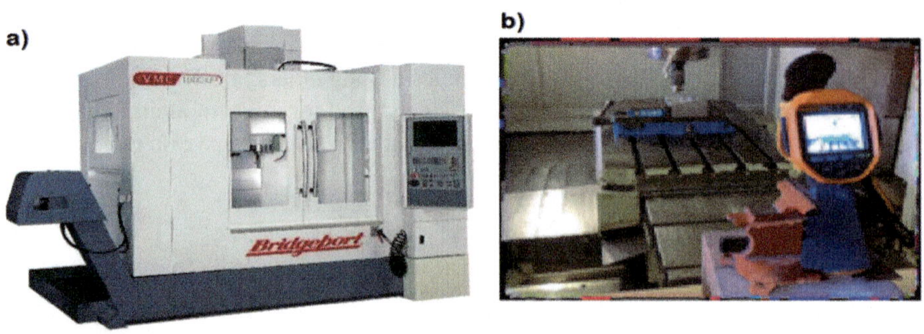

Fig. 2 Equipment used in experimentation. (**a**) Bridgeport high-speed machining center, (**b**) Fluke® Ti400 thermal imager

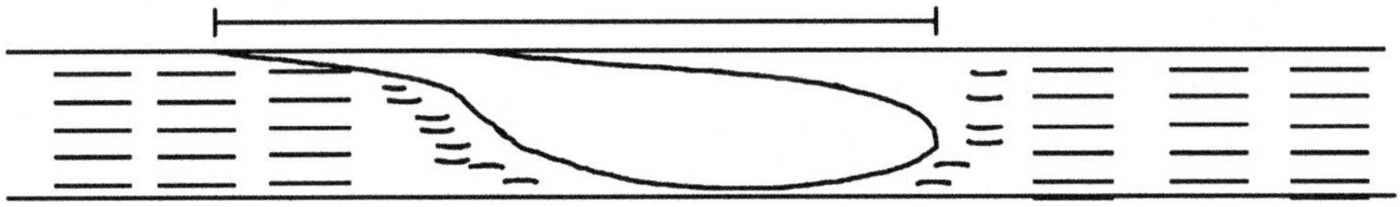

Fig. 3 Schematic of the zones generated in the FSSW process

Table 1 Chemical composition (% e.p.) titanium alloy Ti-6Al-4 V

C	O	N	Fe	Al	V	Ti
0.19	0.19	0.016	0.25	6.2	3.87	Balance

Table 2 Mechanical properties of Ti-6Al-4 V titanium alloy

	Delivery effort (MPa)	Ultimate stress strain (MPa)	HRB hardness
Ti-6Al-4 V	827	896	33

In this regard, having reliable finite element model that is capable of simulating FSSW with minimal possible simulation time can turn out handy to reduce the number of physical experiments required in such studies and applications.

This industry currently uses mechanical bonding technology to join metal components in the manufacture of large, military, and commercial aircraft.

The FSSW process produces a union with different well-defined zones known as the heat-affected zone (HAZ), thermomechanically affected zone (TMAZ), and the agitation zone (ZA) [12], which can be seen in Fig. 3.

Titanium is increasingly important as a material for use in the fuselage because of its excellent fatigue strength, its ability to withstand high temperatures, and its resistance to corrosion [13]; from the industrial point of view, titanium and its alloys have important properties with respect to aluminum and its alloys, steel, and stainless steel. Table 1 shows the chemical composition of the alloy adopted in the numerical model and in Table 2 the mechanical properties of the alloy [6, 14].

During the FSSW process, the operating parameters determine the amount of heat generated for the plastification of the material around the pin, the weld geometry, and thus the mechanical properties of the joint.

2 Experimental Development

The welding process was performed in a high-speed machining center; temperature measurements were successfully performed using a thermal imager (Fluke® Ti400) in the welding zone, capturing the temperature profile at the tool joint and the titanium plates.

During the process, a clamping system was used for the sheets which was placed inside the machining center by positioning the Ti-6Al-4V sheets with an 18 mm

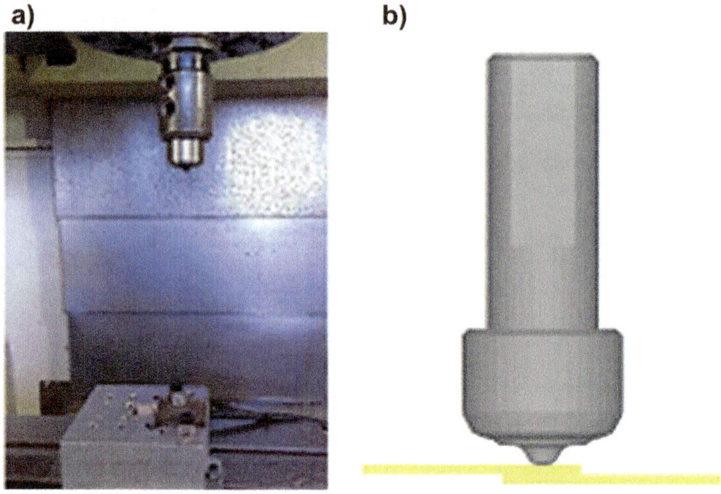

Fig. 4 (**a**) Method of securing the sheets in the machining center. (**b**) Positioning of the tool with respect to the sheets

overlap. The thermographic camera was used by placing it at the side of the work area at a distance of 120 cm. Figure 4 shows the positioning of the tool together with the sheets in the experiment as well as in the simulation.

3 Numerical Modeling

The numerical model was established using the DEFORM-3D finite element software (Fig. 5), in which the parameters shown in Table 3 are considered. The geometry of the tool and the geometries of the plates were accurately replicated based on those used in the experimentation.

Table 3 shows the parameters used in the bonding of Ti-6Al-4V alloy sheets with 2.3 mm thickness.

For the numerical simulation analysis, the Johnson-Cook numerical model was used to predict the behavior of the material when subjected to large deformations, high temperatures, and high deformation rates. The general equation of the Johnson-Cook mathematical model is shown below, and the values used in the simulation are described in Table 4:

$$\sigma = \left[A + B\left(\bar{\varepsilon}^{\mathrm{pl}}\right)^{n}\right]\left[1 + C\ln\left(\frac{\overline{\varepsilon'}^{\mathrm{pl}}}{\overline{\varepsilon'}^{\mathrm{pl}}_{0}}\right)\right]\left[\frac{T - T_{\mathrm{ambient}}}{T_{\mathrm{melt}} - T_{\mathrm{ambient}}}\right] \quad (1)$$

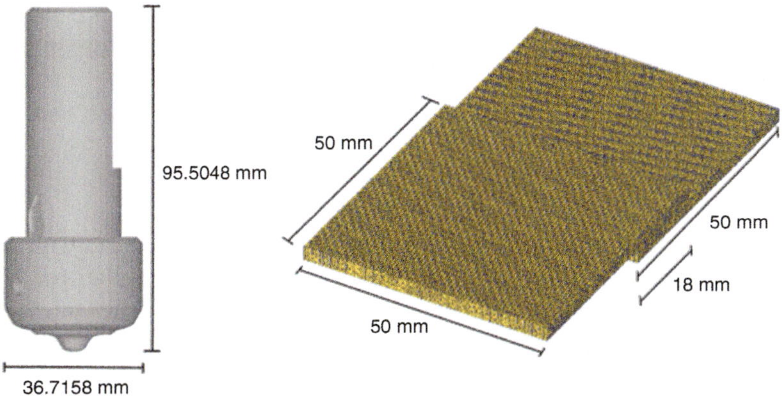

Fig. 5 Dimensions of the plates and the tools

Table 3 Experimental parameters in the FSSW process

Thickness plates	2.3 mm
Dimension	50 × 50 mm
Overlap area	18 mm
Rotational speed (RPM)	500 rpm
Plunge speed (mm/min)	12 mm/min
Temperature	27 °C
Time	3 seg

Table 4 Johnson-Cook equation constants used in the FSSW process simulation

Constant	Symbol	Value
Elastic limit at room temperature (MPa)	A	782.7
Hardening module (MPa)	B	498.4
Coefficient of acrimony or sensitivity to plastic deformation	n	0.28
Coefficient relative to the sensitivity to plastic deformation	C	0.028
Ambient temperature (°C)	T_{ambient}	27
Melting temperature (°C)	T_{melt}	1662
Plastic deformation	$\bar{\varepsilon}^{\mathrm{pl}}$	1
Standard deformation speed	$\bar{\varepsilon'}^{\mathrm{pl}}$	1
Standard deformation speed	$\bar{\varepsilon'}^{\mathrm{pl}}_{0}$	1.00E-.05

Table 5 Characteristics of the developed mesh

Mesh type	Tetrahedral
Number of nodes	31,850
Items	140,028
Absolute interference	Absolute
Method of remalling	Local

In the structured mesh, a refinement was adapted in the contact area with the aim of obtaining precision in the results. The geometry representing the plates of deformable material was placed in the boundary conditions to restrict its movement to represent the actual welding system, and the tool was considered a rigid, nondeformable body. The final mesh developed for the numerical model is shown in Table 5.

4 Discussion

The heating and plastic deformation of the material caused by the rotation of the tool, mainly the shoulder, promote a plasto-viscous behavior in the material. This favors the flow of the tool when the rotary tool moves to the depth of the joint causing a large amount of plastic deformation at the same time as the tool pin generates the flow of the material both in the circumference and in the axial directions. It is considered that this flow mechanism consolidates the bonding of the materials by the extrusion and forging of the material at high rates of deformation. The material flow and the zones generated during the spot welding by friction agitation can be observed by a metallographic analysis. Figure 6 shows micrograph of the cross section and Fig. 7 shows the generated zones.

Figure 7 shows the zones generated during the FSSW process: heat-affected zone (HAZ), thermomechanically affected zone (TMAZ), and stir zone (SZ) as well as the base metal.

On the other hand at two different zone in the center of the tread (agitation zone and heat-affected zone) showed similar results in both zones, creating a uniform weld because there was no significant change between the zones. In the same way the behavior of the temperature was similar in both zone, due to the heat flowed uniformly during the process.

Three points were taken into account in the experimentation as well as in the numerical model to analyze the temperature around the weld trace; in Fig. 8 the evolution of the temperature increase in the FSSW process is shown and Fig. 9 observes evolution of the temperature increase in the simulation of the process.

Figures 8 and 9 show the results of three different points, in which point 1 presents a higher temperature; this is because the reference point taken is the one closest to the footprint in comparison to points 2 and 3 that were taken at a greater distance; this case was presented in both studies.

Figure 10 shows the difference in temperature behavior, which has a higher temperature than experimental; this is due to the points taken as reference since in the simulation and experiment, very similar results were obtained as seen in Fig. 11.

Fig. 6 Micrograph of the cross section of the sample with parameters of 500 RPM and 12 mm/min. mag. 50×

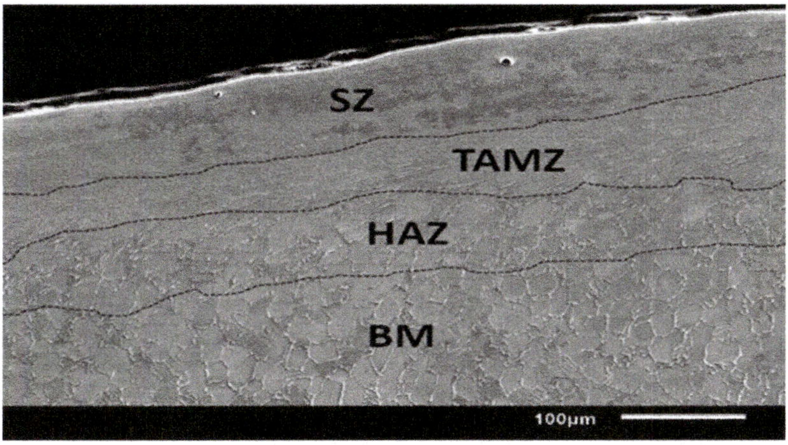

Fig. 7 Zones generated in the process, mag. 100×

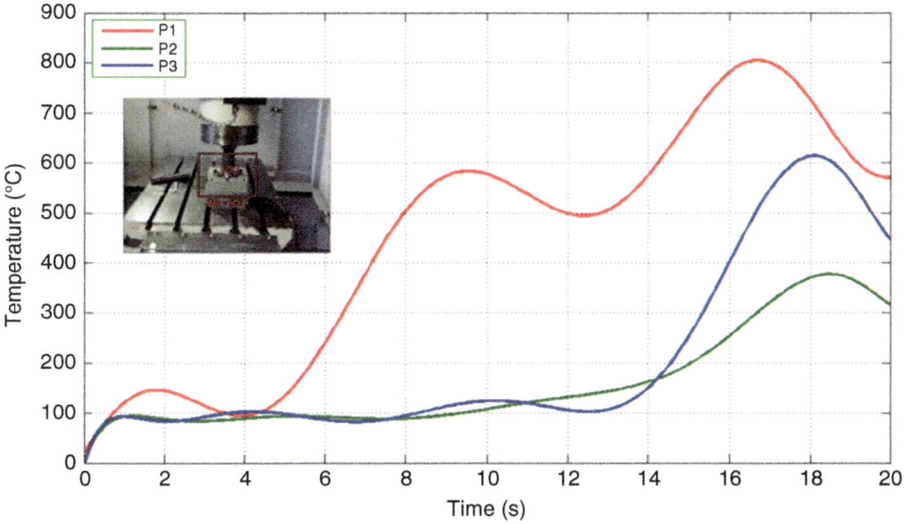

Fig. 8 Evolution of the temperature increase in the FSSW process

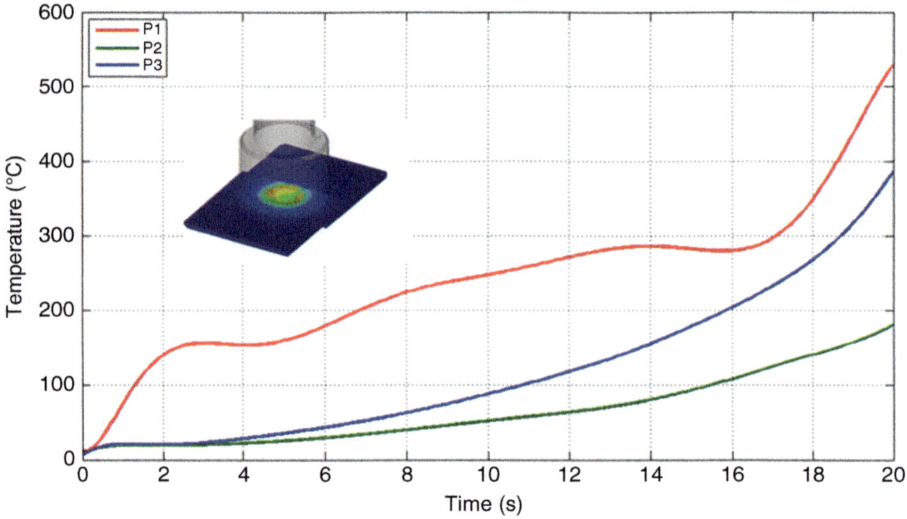

Fig. 9 Evolution of the temperature increase in the process simulation

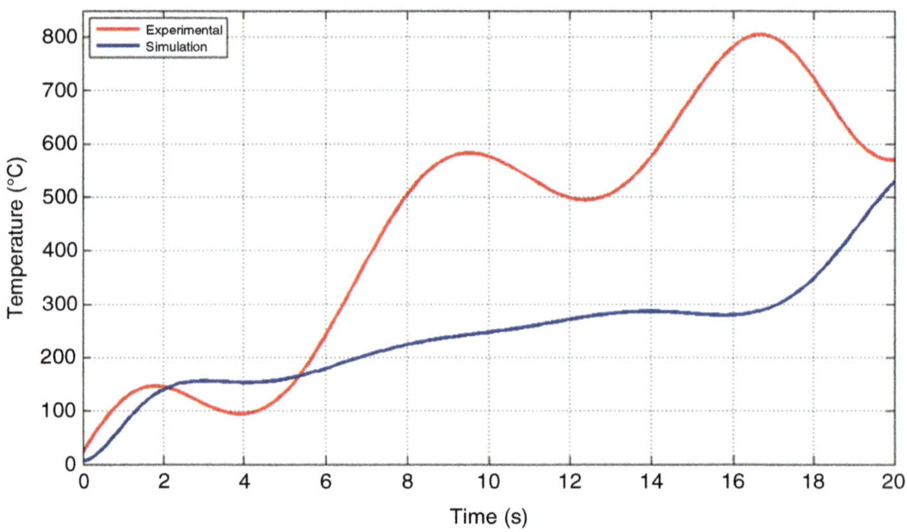

Fig. 10 Comparison of temperature behavior in experimentation and simulation of FSSW process

5 Conclusions

A finite element model of friction stir spot welding (FSSW) was developed for the joining of Ti-6Al-4 V alloy sheets. The tool geometry presented the main effect on material flow and thermal properties.

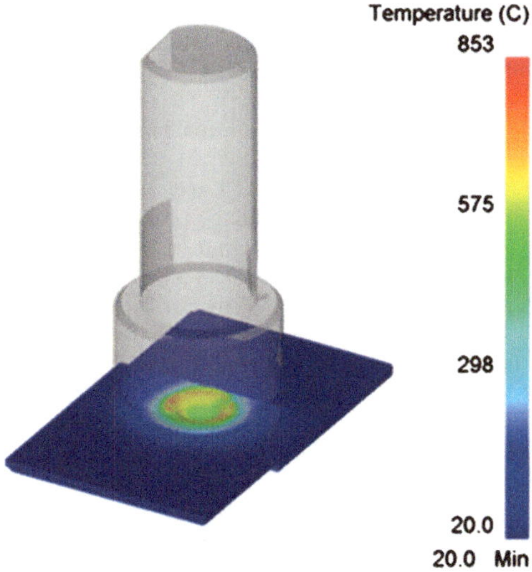

Fig. 11 Temperature distribution in the FSSW simulation

According to the developed study of the finite element model the shape and size of the tool have great influence on the characteristics of the joint. During the process it is subjected to high mechanical stresses at temperatures close to the melting point of the base metal, whereby the material and treatment used for its manufacture are fundamental as regards its useful life.

The parameters values of rotation speed of 500 RPM, plunge speed of 12 mm/s, and 3 s in holding time achieved the joining of the sheets with a thickness of 2.3 mm. The ZA presented fine grains with respect to the MB, while the TMAZ showed the lines of flow that evidence the happened deformation, and the HAZ presented a similar microstructure to the MB.

In the experiment, as well as in the numerical model for analyzing the temperature, the central point on the weld tread was compared, and the maximum temperature reached was obtained. A comparison of the obtained temperature results is shown in Fig. 10 through the FSSW process as well as those obtained in numerical model.

Acknowledgments F. Quiroz is grateful for the financial support granted by PAICYT program. The authors wish to thank the Centro de Investigación e Innovación en Ingeniería Aeronáutica of the Facultad Ingeniería Mecánica y Eléctrica of the Universidad Autónoma de Nuevo León.

References

1. Malafaia, A. M. S., Milan, M. T., Oliveira, M. F., & Spinelli, D. (2010). Fatigue behavior of friction stir spot welding and riveted joints in an al alloy. *Procedia Engineering, 2*(1), 1815–1821. http://dx.doi.org/10.1016/j.proeng.2010.03.195
2. Malik, V., Sanjeev, N. K., Hebbar, H. S., & Kailas, S. V. (2014). Time efficient simulations of plunge and dwell phase of FSW and its significance in FSSW. *Procedia Materials Science, 5*, 630–639. http://dx.doi.org/10.1016/j.mspro.2014.07.309

3. Lathabai, S., Painter, M. J., Cantin, G. M. D., Tyagi, V. K., & Pine Mountain, G. (2006). Friction stir spot welding of automotive lightweight alloys. In *Proceeding go the 7th international conference on trends in welding research*. Georgia: Pine Mountain.
4. Dawes, C. J., & Thomas, W. M. (1996). Friction stir process welds aluminium alloys: The process produces low-distortion, high-quality, low-cost welds on aluminium. *Welding Journal, 75*(3), 41–45.
5. Zimmer, S., Langlois, L., Laye, J., & Bigot, R. (2010). Experimental investigation of the influence of the FSW plunge processing parameters on the maximum generated force and torque. *International Journal of Advanced Manufacturing Technology, 47*(1–4), 201–215. doi:10.1007/s00170-009-2188-3
6. McNelley, T. R. (2010). Procesado por fricción batida (FSP): afino de la microestructura y mejora de propiedades. *Revista de Metalurgia, 46*(Extra), 149–156. http://dx.doi.org/10.3989/revmetalmadrid.19XIIPMS
7. Lippold, J. C., & Livingston, J. J. (2013). Microstructure evolution during friction stir processing and hot torsion simulation of Ti-6Al-4V. *Metallurgical and Materials Transactions A, Physical Metallurgy and Materials Science, 44*(8), 3815–3825. http://dx.doi.org/10.1007/s11661-013-1764-1
8. Buffa, G., Fratini, L., & Micari, F. (2012). Mechanical and microstructural properties prediction by artificial neural networks in FSW processes of dual phase titanium alloys. *Journal of Manufacturing Processes, 14*(3), 289–296. http://dx.doi.org/ 10.1016/j.jmapro.2011.10.007.
9. Khairuddin, J. T., Almanar, I. P., Abdullah, J., & Hussain, Z. (2012). *Principles and Thermo-mechanical model of friction stir welding*. INTECH Open Access Publisher. http://dx.doi.org/10.5772/50156
10. Mishra, R. S., & Ma, Z. Y. (2005). Friction stir welding and processing. *Materials Science and Engineering: R: Reports, 50*(1), 1–78. http://dx.doi.org/10.1016/j.mser.2005.07.001
11. Lee, J. J., Lukachko, S. P., Waitz, I. A., & Schafer, A. (2001). Historical and future trends in aircraft performance, cost, and emissions. *Annual Review of Energy and the Environment, 26*(1), 167–200. doi:10.1146/annurev.energy.26.1.167
12. Tutar, M., Aydin, H., Yuce, C., Yavuz, N., & Bayram, A. (2014). The optimisation of process parameters for friction stir spot-welded AA3003-H12 aluminium alloy using a Taguchi orthogonal array. *Materials & Design, 63*, 789–797. http://dx.doi.org/10.1016/j.matdes.2014.07.003
13. Boyer, R. R. (1995). Titanium for aerospace: Rationale and applications. *Advanced Performance Materials, 2*(4), 349–368. doi:10.1007/BF00705316
14. Arcella, F. G., & Froes, F. H. (2000). Producing titanium aerospace components from powder using laser forming. *JOM Journal of the Minerals, Metals and Materials Society, 52*(5), 28–30. doi:10.1007/s11837-000-0028-x

Parametric Effects in Hybrid Lap Joints of Composite Materials Used in Aircraft Structures

Nabil Vallejo, L.A. Reyes, G.M. Hernández-Muñoz, and P. Zambrano-Robledo

Abstract Composite materials technology in aircrafts needs improved damage growth predictions and effective repair techniques for damaged structures. Improper joint design may lead to overweight or defective structures. The joining of composite materials has traditionally been achieved by mechanical fastening or adhesive bonding. Hybrid joints are a combination of adhesive bonding and mechanical fastening; these joint systems have attracted the attention in aerospace and general mechanical industries due to its higher strength and stiffness and higher energy absorption prior to failure. In this work a parametric study was performed using a finite element software to determine the mechanical resistance of hybrid and simple lap joints. The parametric study consists in the numerical modeling of different conditions that permit estimates the main characteristics of lap joining. The study performed in this investigation seeks for a better understanding of joining of composite materials.

Keywords Fem • Joint strength • Composite materials • Bonded • Hybrid joint

1 Introduction

Aeronautical industry invests a lot of time and money to improve the manufacturing operations, materials, and technologies to guarantee the safety of passengers and crew and at the same time satisfy expectations of the present society. Composite

N. Vallejo • L.A. Reyes (✉) • G.M. Hernández-Muñoz • P. Zambrano-Robledo
Universidad Autónoma de Nuevo León, Facultad de Ingeniería Mecánica y Eléctrica, Centro de Investigación e Innovación en Ingeniería Aeronáutica, México, Ave. Universidad S/N, Cd. Universitaria, San Nicolás de los Garza, NL C.P. 66451, Mexico

Centro de Investigación e Innovación en Insgeniería Aeronáutica, Universidad Autónoma de Nuevo León, México, Aeropuerto Internacional del Norte, Carr. A Salinas Victoria Km. 2.3, Apodaca, NL C.P. 66600, Mexico
e-mail: luis.reyessr@uanl.edu.mx

P. Zambrano-Robledo et al. (eds.), *Proceedings of the Symposium of Aeronautical and Aerospace Processes, Materials and Industrial Applications*,
DOI 10.1007/978-3-319-65611-3_12

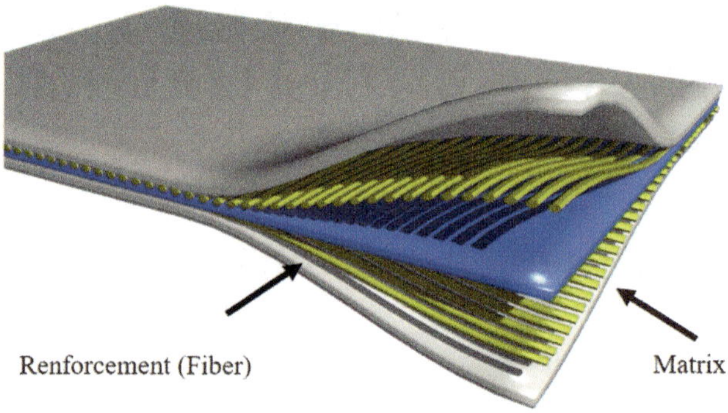

Fig. 1 Structure of composite materials, reinforcement, and matrix [3]

materials are some of the new trends in the aeronautical sector, as well as in many other industries, for example, automotive structures, recreational and sporting equipment, industrial process, and many others. The impact of composite materials is due to the number of advantages compared to conventional engineering materials, such as aluminum and steel. The main characteristics of composite materials are the combination of low density associated with high stiffness and strength, which are required for materials that make up the aeronautical structures. Composite materials consist of two or more materials combined together (physically or chemically distinct) to obtain a new alloy with enhanced properties [1, 2]. These materials contain strong fibers, continuous or noncontinuous, that are inlaid in a weaker material called matrix. The function of the matrix is to keep the geometric arrangement of fibers and transmit the load acting on the composite component. The fibers consist of several hundred or thousands of filaments; one of the most common fibers used in aeronautic is the carbon fiber. The composite structure is arranged during manufacture of components with fibers orientated in various directions in sufficient concentrations to provide the required strength and optimally distribute the stresses; see Fig. 1 [3]. Composite materials have emerge due to the necessity to improve and obtain engineering materials able to cover or satisfy the demand of new technologies and capabilities in diverse industries, i.e., aeronautical, automotive, mechanical, medical, power, petrochemical, etc. Some of the main characteristics of composites that have made them such popular materials are lightness (which is one of the main reasons why they are highly used in aeronautical structures), flexibility, high corrosion resistance, impact strength, and fatigue strength [4, 5].

Mechanical joints are classified in three types: bolted, bonded (adhesive), and hybrid joints, the last one being a combination of the first two techniques—bolted and bonded joints. Adhesive bonding joints avoid the use of the fasteners and fastener holes, which promote stress concentrations; this type of joints produces the assembly through an adhesive, mostly epoxy. However, adhesive materials are susceptible to degradation under environmental conditions. Hybrid joints are a combination of adhesive bonding and mechanical fastening joints and have attracted the attention of aerospace sector and general mechanical industries, due to its advantages in comparison to simple joining techniques such as higher strength and stiffness and higher energy absorption prior to failure. Parametric studies allow the evaluation of mechanical effects when the main parameters are changed; these parameters can be dimensional and material related.

A previous parametric study was developed by Nagesh Ganji [1] to analyze the effects of different parameters such as material properties, tensile loading, adherent thickness, bolt diameter, and overlap length on the load transfer by a bolt. It was noticed that the hybrid joint can provide enhanced structural performance when compared to adhesive joint. The study also discusses the modeling of contact between the bolt and the hole and nonlinear material behavior of the model due to defective bonded joint. Sailaja Malladi [6] performed a parametric analysis of a double-lap joint, revealing the effects of several factors on the structural performance of the analyzed joint. In [7], Hart Smith performed an investigation of bonded-bolted composite joints. The design and analysis of high-load-transfer joints in advanced composite structures were made in Fortran. Analysis of undamaged structures showed that due to adhesive bond, load path is so much stiffer than the load path through bolts or rivets; the combination is no stronger than a well-designed bonded joint alone. However, the combination of bonding and bolting is shown to be particularly useful for repair and to prevent damage from spreading.

The present study investigates the parametric effects in a hybrid lap joint through numerical simulations to determine the load transfer and provide information about what model or under what characteristics the results are more efficient or less efficient, according to the establish objective of this investigation.

2 Finite Element Method

The most common technique for obtaining numerical solutions to differential equations and nontrivial domains are the finite element methods, where the given domain is subdivided into simple geometry objects, called elements, and an approximate solution is computed as a linear combination of basic functions [8–10]. One of the main characteristics of FEM is that continuum field, well known as domain, is subdivided into cells, called elements, which form a grid. The two-dimensional elements have a triangular or quadrilateral form, and this one can be rectilinear or curved, while the grid does not need to be structured. When the grid is unstructured and the cells or elements are curved, the complex geometries can be easily handled.

In this work a FEM was developed in ANSYS APDL; Shell181 element was used for the composite plate analysis, Solid185 element was selected for the rivet, and an initial pretension was considered. Solid model characteristics are shown in Fig. 2 and Table 1. The total number of nodes and elements defined were 6256 and 28,349, respectively.

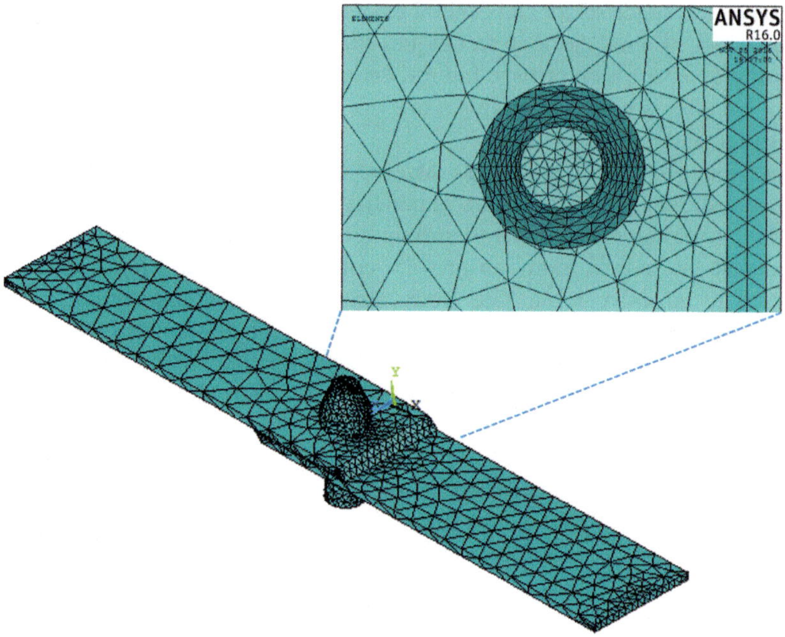

Fig. 2 Finite element model of hybrid lap joint

Table 1 Materials and parameters defined in the numerical models

Model	Material	Adhesive thickness (mm)
1	Carbon 3K plane fiber/ Epoxy	1.8
2	Carbon 3K plane fiber/ Epoxy	3.6
3	Carbon 3K plane fiber/ Epoxy	5.4
4	Carbon Fiber/ Epoxy USN150	1.8
5	Carbon Fiber/ Epoxy USN150	3.6
6	Carbon Fiber/ Epoxy USN150	5.4
7	Glass Fiber/ Epoxy UGN150	1.8
8	Glass Fiber/ Epoxy UGN150	3.6
9	Glass Fiber/ Epoxy UGN150	5.4

3 Methodology

The work is divided in two parts; first part of the study consists in performing a parametric analysis by FEM, in which the recreation of the proposed joints used in aeronautics will be made with ANSYS software. The main focus consists in the modeling of defined samples according to the ASTM D 3479/D 3479 M, Standard Test Method for Tension-Tension Fatigue of Polymer Matrix Composite Materials [11]. A sketch of dimensions of samples can be observed in Fig. 3.

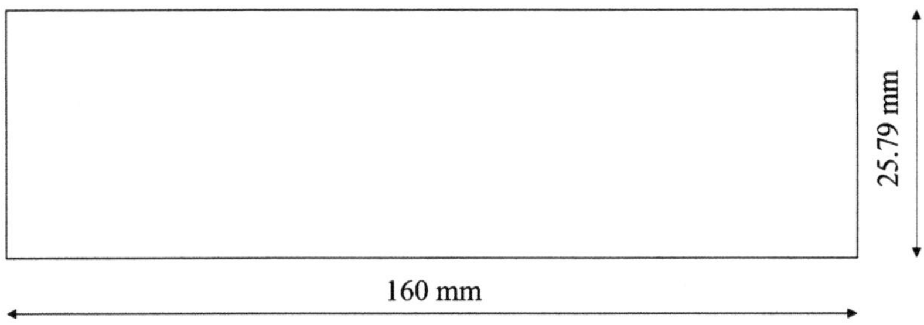

Fig. 3 Dimension of the composite plates

Table 2 Properties of Carbon fiber/Epoxy USN150 [2]

E_X (GPa)	E_Y (GPa)	E_Z (GPa)	G_X (GPa)	G_Y (GPa)	G_Z (GPa)	υ	ρ (kg/m^3)
131	8.2	8.2	4.5	4.5	3.5	0.281	1360

Once the geometry is predefined, the properties of the material have to be established, and the specimens will be specified from three different materials, carbon fiber, aluminum, and glass fiber, following a sequence of $[(0/90)/(45/-45)/(45/-45)/(0/90)]_3$ for the distribution of the layers that will compose the sample; such sequence allows a better distribution of stresses; the number of layers for each plate is 12; once that both plates (upper and lower) are assembled, the sample will represent a lap joint of 24 layers between the upper and lower plate adding the thickness of the adhesive. The principal properties of selected materials of this study can be observed in Tables 2–6.

The value of pretension was defined as 5 kN and the friction coefficient as 0.2. To represent the hybrid lap joints, the geometry of the rivet and the adhesive is included, defining the properties of each one, and subsequently, an arrangement of all the geometries has to be performed to generate the specimen in order to represent the respective parametric model. Finally, the proposed configuration will be submitted to constant pressure in the X-axis (−3000 N) in order to obtain the results of total distribution in X component, Von Mises stress, and shear stress in XZ component for different parametric models shown in Table 1.

The simulation of the different models proposed to obtain the parametric effects of the hybrid lap joint was performed in ANSYS Mechanical APDL. The CAD of the rivet, adhesive, and plates to previously assemble the hybrid joint was made in SolidWorks software, Fig. 4.

The overlap of the samples to recreate hybrid joint configuration was taken as 25 mm between one and the other, according to a previous analysis [6] that proves a better resistance to fatigue.

Material properties for carbon and glass fiber were collected from [2]; see Table 2. The carbon/glass fibers were developed at a curing time of 24 h and orientation $[(0/90)/(45/-45)/(45/-45)/(0/90)]_3$. Adhesive (epoxy) and rivet properties were

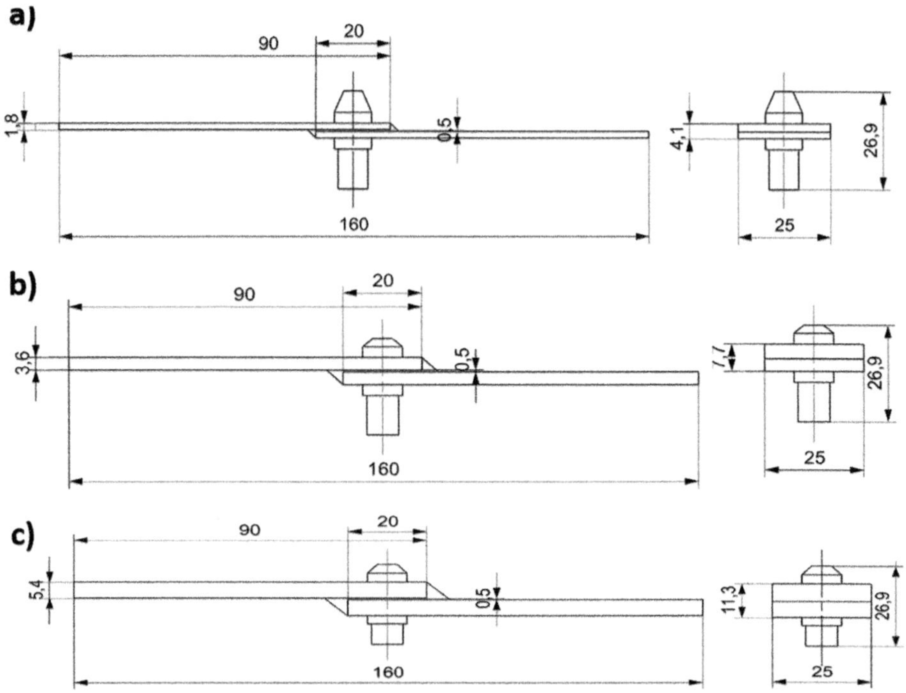

Fig. 4 Hybrid joint configuration, (**a**) adhesive thickness of 1.8 mm, (**b**) 3.6 mm, and (**c**) 5.4 mm

Table 3 Properties of Carbon 3K plane fiber/Epoxy [3]

E_X (GPa)	E_Y (GPa)	E_Z (GPa)	G_X (GPa)	G_Y (GPa)	G_Z (GPa)	υ	ρ (kg/m^3)
98	7.8	7.8	4.7	4.7	3.2	0.34	3000

Table 4 Properties of Glass fiber/Epoxy UGN150 [4]

E_X (GPa)	E_Y (GPa)	E_Z (GPa)	G_X (GPa)	G_Y (GPa)	G_Z (GPa)	υ	ρ (kg/m^3)
43.5	5	5	5	5	5	0.25	1280

obtained from [9], assuming that carbon/glass fibers are orthotropic materials and adhesive and aluminum alloy rivet are isotropic.

The second part of the work is dedicated to the analysis of acquired data, comparing the results obtained from finite element simulations with different parameters. After obtaining numerical results for all model conditions, it will be discussed how the parameters, materials, and thickness influence the behavior of the hybrid lap joint configuration.

Table 5 Properties of film adhesive material /Epoxy

E (MPa)	υ	ρ (kg/m^3)
3400	0.4	1750

Table 6 Properties of aluminum alloy rivet 2017-T4 [9]

E (GPa)	υ	ρ (kg/m^3)
72.4	0.33	2790

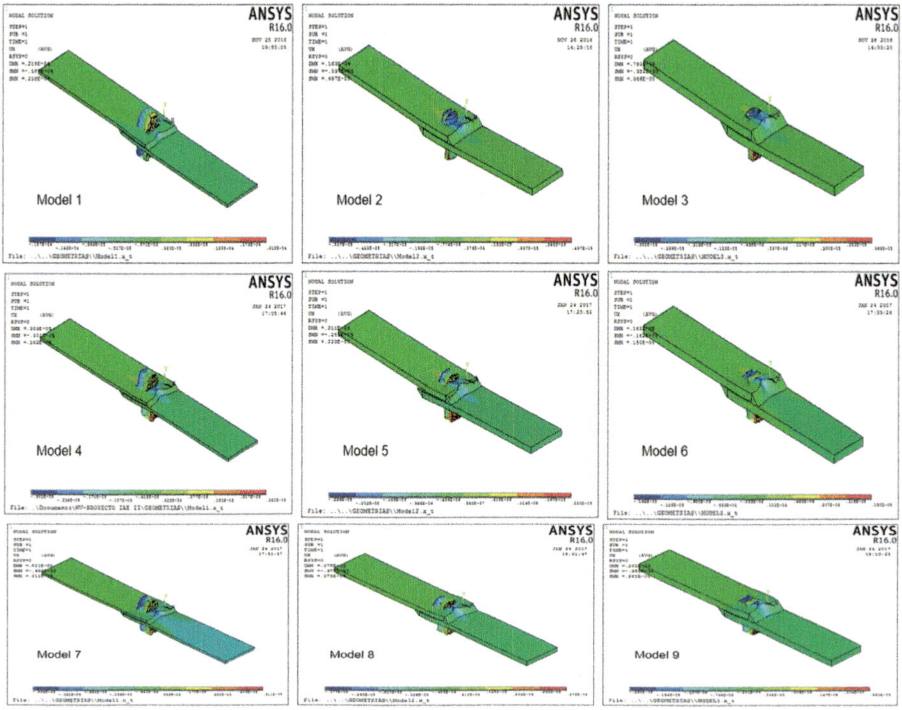

Fig. 5 Numerical results of displacement in X component for nine hybrid lap joint models evaluated

4 Results and Discussion

Numerical results of displacement during loading are observed in Fig. 5; deformation is scale-up to see deformation characteristics; a defined behavior on deformations of the plates is observed due to the load transfer of the joint system.

Figure 6 presents the numerical result of displacement in component *x* for the models developed; the larger displacement is found in model 1 of Carbon 3K plane fiber and the lower displacement is present in model 7 of glass/fiber epoxy with a

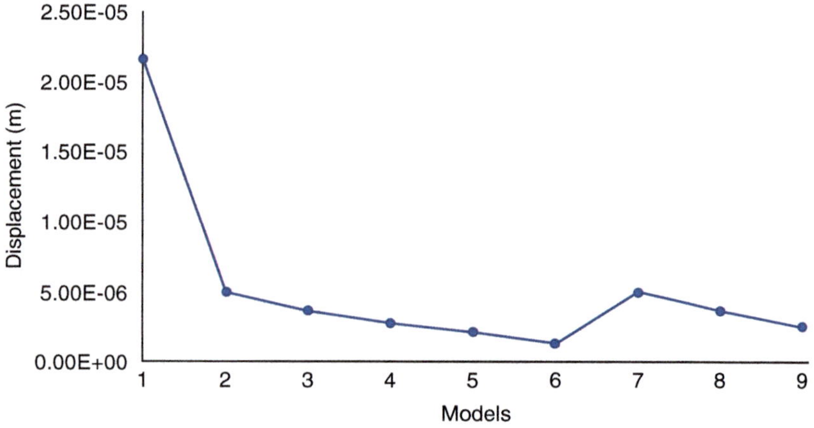

Fig. 6 Displacement in X component for nine hybrid lap joint models evaluated

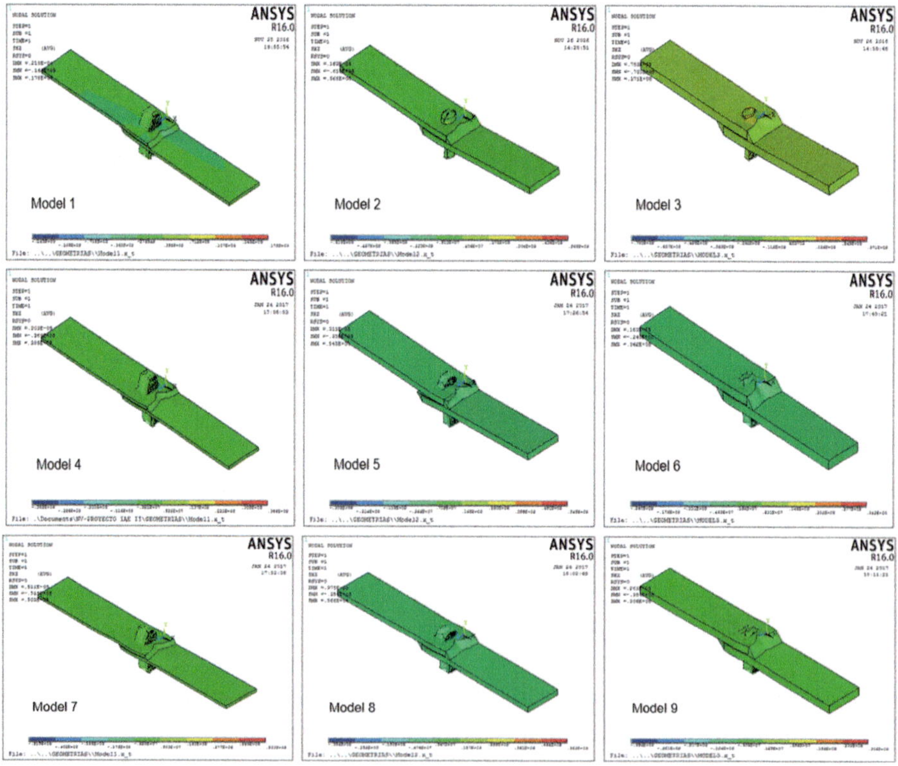

Fig. 7 Numerical simulation results of shear stress in XZ axis for the nine hybrid lap joint models

value of 0.00511 mm. The shear stress in *XZ* is observed in 9 models (Fig. 7), the values are scale-up to see the deformation pattern during loading, and Fig. 8 shows the numerical results of maximum shear stress in XZ component. Higher values are found in models 7 and 8 with glass/fiber epoxy of 1.8 mm and 3.6 mm, respectively.

The Von Mises stress values are depicted in Fig. 9. Higher values are observed in models 1 and 2 made with Carbon 3K plane fiber; model 3 and model 6 have the same adhesive thickness of 5.4 mm; however, the Von Mises stress is about 40% lower for model 6 consisting of Carbon fiber/Epoxy USN150. Figure 10 presents a contour plot of the nine lap joint numerical models developed. From this figure it can be observed that higher shear stress values are present from models 8–9 made of glass/fiber epoxy with thickness of 3.6 and 5.4 mm, respectively. Part b of Fig. 10 depicted the Von Mises behavior of nine models developed; models from 1 to 3

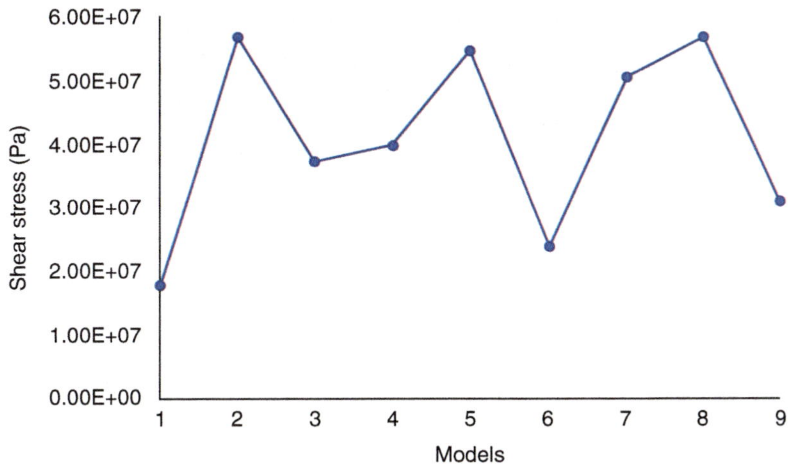

Fig. 8 Shear stress in XZ component for the 9 hybrid lap joint models

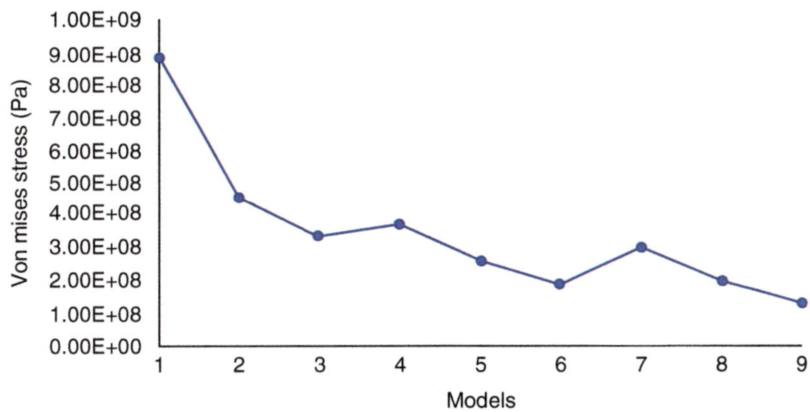

Fig. 9 Von Mises stress graphic in XZ component for the nine hybrid ap joint models

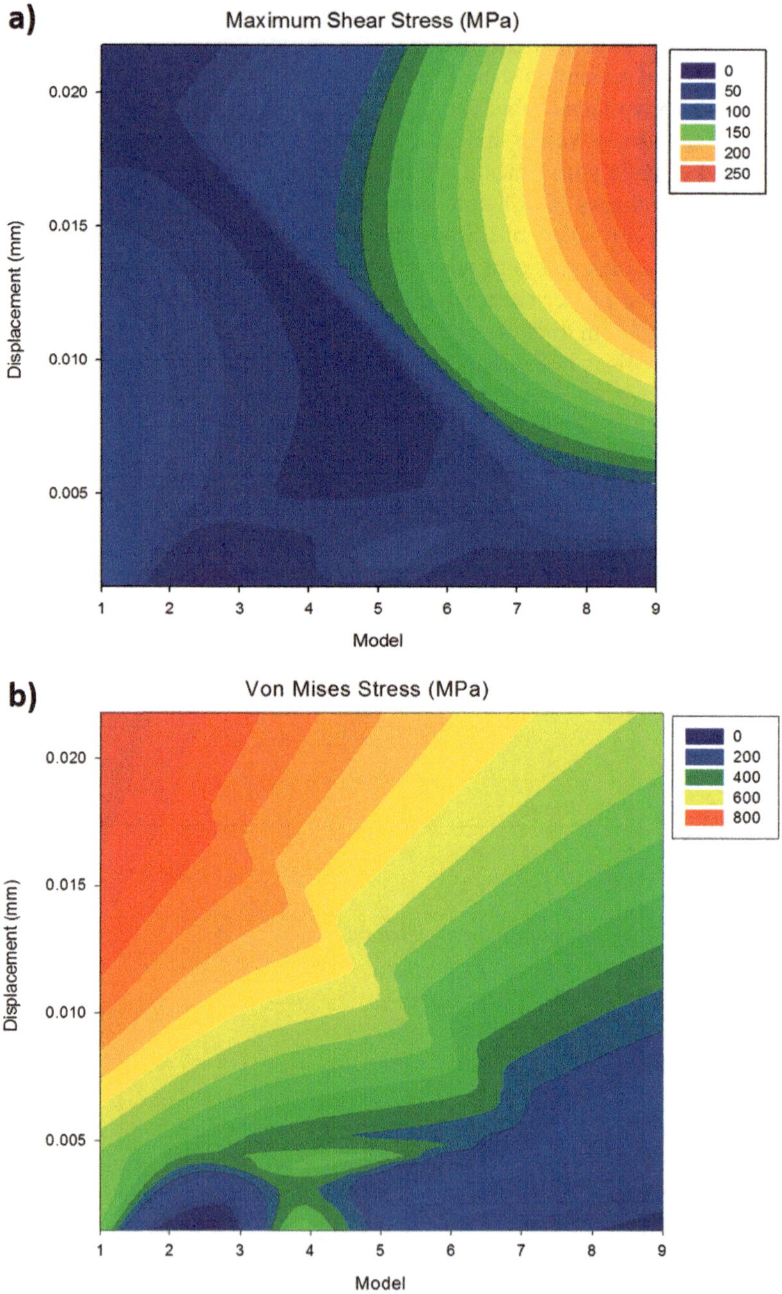

Fig. 10 Contour plots of lap joint numerical models, (**a**) shear stress in XZ component, and (**b**) Von Mises stress

present higher values of stress with displacements above 0.010 mm. From this figure it is evident that shear stresses can be reduced considerably with Carbon 3K plane fiber and Carbon fiber/Epoxy USN150 up to 3.6 mm of thickness; however, lower Von Mises stresses are obtained from carbon and glass/fiber epoxy.

5 Conclusions

A numerical model was developed to understand the performance of hybrid lap joints. The variation of different characteristics, such as materials, geometries, and adhesive thickness, was evaluated in finding the maximum deformation in the main axis, the maximum shear stress, and the Von Mises stress. It was found a maximum shear stress in model 2 with a value of 56.8 MPa and a maximum Von Mises stresses in the joint was present in model 1 with a value of 886 MPa, while the lower value was present in model 9 with 131 MPa. Shear stresses can be reduced considerably with Carbon 3K plane fiber/Epoxy and Carbon fiber/Epoxy USN150 up to 3.6 mm of thickness. In agreement with previous studies, models with bigger adhesive thickness present poor resistance compared to that of a small thickness because of the use of adhesive in bigger quantities representing uncertainty to predict its modes of failure due to their complexity and to the fact that this type of materials are nonlinear.

References

1. Ganji, N. (2001). *Parametric study of load transfer in two-bolted single lap hybrid (bonded/ bolted) shear joints*. Hyderabad, India: B.E. Chaithanya Bharathi Institute of Technology.
2. Qin, M. (December 2003). *Theoretical and experimental analysis of adhesive composite joints*. Lincoln, Nebraska: The Graduate Collage at the University of Nebraska.
3. Gay, D. (2014). *Composite materials, designs and applications* (3rd ed.). France: CRC Press, Laboratoire de Génie Mécanique de Toulouse. ISBN: 978-1-4665-8487-7.
4. Kar, K. K. (2016). *Composite materials, processing, applications, characterizations*. Kanpur: Advance Nano engineering. Materials Laboratory, Indian Institute of Technology Kanpur.
5. Gonzalez, V., & Barcia, F. L. (2006). *Composite materials based on modified epoxy resin and carbon fiber*. Instituto de Macromoléculas, Centro de Tecnología, Universidad Federal de Rio de Janeiro: Brazil.
6. Malladi, S. (2004). *Parametric modeling and analysis of structural bonded joints*. Morgantown, WV: College of Engineering and Mineral Resources at West Virginia University, Department of Mechanical and Aerospace Engineering.
7. Hart Smith, L. J. (1985). *Bonded-bolted composite joints*. Long Beach: Douglas Aircraft Company.
8. Öchsner, A. (2016). *Computational statics and dynamics, an introduction based on the finite element method*. Southport: Griffith School of Engineering. ISBN: 978-981-10-0732-3.
9. Pillwein, V. (2010). *An introduction to finite element methods*. Linz: Johannes Kepler University.
10. Wendt, J. F. (2009). *Computational fluid dynamics, an introduction. Director von Karman Institute for Fluid Dynamics*. Belgium: Sint-Genesius-Rode. ISBN: 978-3-540-85055-7.
11. American Society for Testing and Materials. (1996). *ASTM D 3479/D 3479M, Standard test method for tension-tension fatigue of polymer matrix composite materials*. Philadelphia: ASTM.

Index

P. Zambrano-Robledo et al. (eds.), *Proceedings of the Symposium of Aeronautical and Aerospace Processes, Materials and Industrial Applications*,
DOI 10.1007/978-3-319-65611-3

Printed by Printforce, the Netherlands